The Geology of Canyonlands National Park

By William A. Szary

Copyright 2021, Earth2Energy. All Rights Reserved.

Book Cover: Canyonlands National Park viewed from the Green River. Rico Formation is at river level beneath the Cedar Mesa Sandstone member of the Cutler Formation. Overlying the Cedar Mesa Sandstone is the red sloping talus forming Moenkopi formation. The cliff forming Chinle Formation is positioned on top. Source: Ron Niebrugge Photography posted on the internet.

Library of Congress Catalog in Publications Data:

Szary, William A. The Geology of Canyonlands National Park

Includes references

ISBN 13: 9798507858538

Earth2Energy Educational Publishing
Port Richey FL 34668

Earth2Energy is a Registered Trademark

Table of Contents

Chapter 1. Introduction to Canyonlands National Park 5
Physiography
Geological Overview

Chapter 2. Canyonlands Stratigraphy Overview 10
Hermosa Group

The Cutler Group
White Rim Sandstone
Organ Rock Shale
Cedar Mesa Sandstone
Elephant Canyon Formation
Halgaita Shale

Moenkopi Formation

Chinle Formation
Stratigraphy of the Chinle Formation

The Glen Canyon Group
The Wingate Formation
The Kayenta Formation

Navajo Sandstone
Iron Oxide Concretions

The Entrada Sandstone

Chapter 3. Uplifts 36
The Laramide Orogeny

Chapter 4. Canyonlands National Park Field Tour 39
Geographic Setting
Rocks and Landforms

Park Observations
The High Mesas
Island in the Sky
Dead Horse Point State Park
North Entrance
Shafer White Rim Trail
Grand View Point

Green River Overlook
Upheaval Dome
Hatch Point
Canyonlands Overlook
U-3 Loop
Anticline Overlook
Orange Cliffs

The Bench-land
The Maze and Land of Standing Rocks
The Needles District
Salt, Davis, and Lavender Canyons
The Needles and the Grabens

Canyons of the Green and Colorado Rivers
Entrenched and Cut Off Meanders
The Colorado River

Chapter 5. Geologic History Summary 104

References 107

Chapter 1. Introduction to Canyonlands National Park

Wikipedia (2021) posted an introduction to Canyonlands National Park on the internet which is summarized in this section.

Canyonlands National Park is an American national park located in southeastern Utah near the town of Moab. The park preserves a colorful landscape eroded into numerous canyons, mesas, and buttes by the Colorado River, the Green River, and their respective tributaries. The park is divided into four districts: the Island in the Sky, the Needles, the Maze, and the combined rivers—the Green and Colorado—which carved two large canyons into the Colorado Plateau. While these areas share a primitive desert atmosphere, each retains its own character (**Figure 1**).

Figure 1. Regional satellite image of Canyonlands National Park, Utah. The major districts are labeled: The Island in the Sky, The Maze, and the Needles.

Physiography

The Colorado River and Green River combine within the park dividing it into three districts called the Island in the Sky, the Needles, and the Maze. The Colorado River flows through Cataract Canyon below its confluence with the Green River (**Figure 2**). The Island in the Sky district is a broad and level mesa in the northern section of the park between the Colorado and Green Rivers.

The district has many viewpoints overlooking the White Rim, a sandstone bench 1,200 feet (370 m) below the Island, and the rivers which are another 1,000 feet (300 m) below the White Rim.

Figure 2. Cataract Canyon occurs below the confluence of the Green and Colorado Rivers. It is the length of the entire canyon. The big bend location is in the center of the canyon at the red marker.

The Needles district is located south of the Island in the Sky on the east side of the Colorado River. The district is named for the red and white banded rock pinnacles which are a major feature of the area. Various other naturally sculpted rock formations are also within this district including grabens, potholes, and arches. Unlike Arches National Park where many arches are accessible by short to moderate hikes, most of the arches in the Needles district lie in backcountry canyons requiring long hikes or four-wheel drive trips to reach them (**Figure 3**).

The Ancestral Puebloans inhabited this area and some of their stone and mud dwellings are well-preserved, although the items and tools they used were mostly removed by looters. The Ancestral Puebloans also created rock art in the form of petroglyphs most notably on Newspaper Rock along the Needles access road. The Maze district is located west of the Colorado and Green rivers. The Maze is the least accessible section of the park, and one of the most remote and inaccessible areas of the United States (**Figure 4**).

A geographically detached section of the park located north of the Maze district, Horseshoe Canyon contains panels of rock art made by hunter-gatherers from the Late Archaic Period (2000-1000 BC) pre-dating the Ancestral Puebloans. Originally called Barrier Canyon, Horseshoe's artifacts, dwellings, pictographs, and murals are some of the oldest in America. The images depicting horses date from after 1540 AD when the Spanish reintroduced horses to America.

Figure 3. The Needles displays rock fracturing patterns oriented towards the northeast. The actual bedding planes are horizontally oriented.

Since the 1950s, scientists have been studying an area of 200 acres (81 ha) completely surrounded by cliffs. The cliffs have prevented cattle from over grazing on the area's 62 acres (25 ha) of grassland. According to the scientists, the site may contain the largest undisturbed grassland in the Four Corners region. Studies have continued biannually since the mid-1990s. The area has been closed to the public since 1993 to maintain the nearly pristine environment.

Geologic Overview

A subsiding basin and nearby uplifting mountain range (the Uncompahgre) existed in the area in Pennsylvanian time. Seawater trapped in the subsiding basin created thick evaporite deposits by Mid Pennsylvanian. This, along with eroded material from the nearby mountain range became the Paradox Formation, itself a part of the Hermosa Group. Paradox salt beds started to flow later in the Pennsylvanian and probably continued to move until the end of the Jurassic.

Some scientists believe Upheaval Dome was created from Paradox salt bed movement creating a salt dome, but more modern studies show that the meteorite theory is more likely to be correct.

Figure 4. The Maze is remote, carved by dendritic drainage patterns leaving ridges mimicking the drainage, Ridges represent inverted topography resulting from the erosion.

A warm shallow sea again flooded the region near the end of the Pennsylvanian. Fossil-rich limestones, sandstones, and shales of the gray-colored Honaker Trail Formation resulted (**Figure 5**).

A period of erosion then ensued creating a break in the geologic record called an unconformity. Early in the Permian, an advancing sea laid down the Halgaito Shale Coastal lowlands later returned to the area, forming the Elephant Canyon Formation. Large alluvial fans filled the basin where it met the Uncompahgre Mountains creating the Cutler red beds of iron-rich arkose sandstone. Underwater sand bars and sand dunes on the coast inter-fingered with the red beds and later became the white-colored cliff-forming Cedar Mesa Sandstone. Brightly colored oxidized muds were then deposited forming the Organ Rock Shale. Coastal sand dunes and marine sand bars once again became dominant, creating the White Rim Sandstone. A second unconformity was created after the Permian sea retreated. Flood plains on an expansive lowland covered the eroded surface and mud built up in tidal flats creating the Moenkopi Formation. Erosion returned forming a third unconformity. The Chinle Formation was then laid down on top of this eroded surface.

Figure 5. The Honaker Trail Formation is exposed in Canyonlands National Park along the Colorado River. Source: National Park Service posted on the internet.

Increasingly dry climates dominated the Triassic. Therefore, sand in the form of sand dunes invaded and became the Wingate Sandstone. For a time climatic conditions became wetter and streams cut channels through the sand dunes forming the Kayenta Formation. Arid conditions returned to the region with a vengeance. A large desert spread over much of western North America and later became the Navajo Sandstone. A fourth unconformity was created by a period of erosion.

Mud flats returned forming the Carmel Formation (**Figure 6**). The Entrada Sandstone was laid down next. A long period of erosion stripped away most of the San Rafael Group in the area along with any formations that may have been laid down in the Cretaceous period. The Laramide orogeny started to uplift the Rocky Mountains 70 million years ago and with it the Canyonlands region. Erosion intensified and when the Colorado River Canyon reached the salt beds of the Paradox Formation the overlying strata extended toward the river canyon forming features such as The Grabens. Increased precipitation during the ice ages of the Pleistocene quickened the rate of canyon excavation along with other erosion. Similar types of erosion are ongoing, but occur at a slower rate.

Chapter 2. Canyonlands Stratigraphy Overview

The stratigraphic column discussion presents formational units from oldest to youngest. The exposed geology of the Canyonlands area is complex and diverse. Twelve (12) formations are exposed in Canyonlands National Park that range in age from Pennsylvanian to Cretaceous. The oldest and perhaps most interesting was created from evaporites deposited from evaporating seawater. Various fossil-rich limestones, sandstones, and shales were deposited by advancing and retreating warm shallow seas through much of the remaining Paleozoic.

Figure 6. The Carmel Formation appears in Canyonlands as a non-descript mud flat deposit covered by Entrada Sandstone (white). Source: National Park Service posted on the internet.

Eroded sediment from a nearby mountain range later mixed with coastal dune and sand bar deposits. The end of the Paleozoic and the start of the Mesozoic saw the last seas start to leave the region for good. A subdued topography was dominated by flood plains and tidal flats. Much further inland, the Triassic climate in the region was dry. Vast deserts covered much of that part of North America except for one period when streams for a time fought the sand dunes. Wetter times returned. The uplifting of the Rocky Mountains starting in late Cretaceous greatly affected the Canyonlands region. Erosion rates increased and further quickened the onset of the ice ages in the Pleistocene. Modern-day erosion occurs at a slower rate.

Hermosa Group

A vast sea covered the region in early Pennsylvanian time. A basin in the area called Paradox Basin subsided and a mountain range called the Uncompahgre Mountains was uplifted to the east. Great quantities of seawater were trapped in the subsiding basin and water became increasingly saline in the hot and dry climate. Thousands of feet of evaporites (anhydrite and gypsum then halite) started to build up in the Mid Pennsylvanian and storms occasionally washed sediment from the nearby mountains. Fresh seawater periodically refilled the basin but was never able to flush out the very salty water there (the new water in fact floated on top of the brine).

These beds were later lithified to become the Paradox Formation which in turn is part of the Hermosa Group. Compressed salt beds from the Paradox started to flow plastically later in the Pennsylvanian and probably continued to move from then until the end of the Jurassic. Satellite-based measurements indicate that flow of salt and gypsum continues today to cause flexing and faulting of overlying sedimentary layers. The Paradox is up to 5000 feet (1520 m) thick in places and in the park is exposed at the bottom of Cataract Canyon as rock gypsum inter-bedded with black shale (**Figure 7**). Upward movement of the Paradox is also a possible theory for the creation of Upheaval Dome, although none of the Paradox is exposed on the dome, the predominant theory being a meteor crater.

Figure 7. Black shale is intermixed with rock gypsum in the lower part of Cataract Canyon in the right basal cliff face of the Hermosa Group Paradox Formation. A fault disrupts beds in the center right. The black coloration on the left side cliff face belongs to manganese desert varnish above the side channel entering the Colorado River. Source: World Wide River Expeditions.

A warm shallow sea again flooded the region near the end of the Pennsylvanian. Limey oozes, sand, and mud were deposited on top of the salt-filled basin. These sediments became the fossil-rich limestones, sandstones, and shales of the gray-colored Honaker Trail Formation. Outcrops of the Honaker Trail can be seen near the bottom of deep canyons in the park most notably along the Colorado River. A period of erosion then ensued creating a break in the geologic record called an unconformity.

The Cutler Group

Early in the Permian, a transgressing (advancing) sea laid down the Halgaito Shale. Coastal lowlands returned to the area after the sea regressed (retreated) forming the Elephant Canyon Formation. These formations can now be seen in Cataract and Elephant Canyons. The Uncompahgre Mountains (Uncompahgre Plateau) were undergoing extensive erosion during this time. Large alluvial fans filled the basin where it met the range. The resulting Cutler red beds are made of iron rich arkose sandstone. Underwater sand bars and sand dunes on the coast inter-fingered with the red beds and later became the white colored cliff-forming Cedar Mesa Sandstone. Today these two competing rock units are exposed in a 4 to 5 mile (6.4 to 8 km) wide belt across the park stretching from south of the Needles through the Maze and to the Elaterite Basin (**Figure 8**).

Figure 8. Elaterite Basin exposes reddish colored Elephant Canyon Formation at ground level, part of the Cutler Group. White sandstone belonging to the Cedar Mesa Formation caps the red beds. Organ Rock Shale covered the Cedar Mesa Formation and the White Cliff Sandstone is the flat plateau landform. The Elaterite Butte is the prominent butte in the right center on top of the rock stack. The location is in the Maze District of the park. Source: Flickr posted on the internet.

Brightly colored oxidized muds were deposited on top of the Cedar Mesa and ranged in color from red to brown. These sediments eventually became the slope-forming Organ Rock Shale formation and can be seen in the Land of Standing Rocks part of the park. Coastal sand dunes and marine sand bars once again became dominant creating the cross-bedded cliff-forming White Rim Sandstone. It is exposed as a topographic bench 1200 feet (365 m) below the top of Island in the Sky (thus earning its name) and along the White Rim Road. A fossilized offshore sand bar made of the White Cliff Sandstone is also exposed in the Elaterite Basin. A tarry dark-brown oil called elaterite seeps out of the structure giving the basin its name. The Permian sea retreated which exposed the land to a long period of erosion and thus created a second unconformity.

The Cutler Formation or Cutler Group is a rock unit that is spread across the U.S. states of Arizona, northwest New Mexico, southeast Utah and southwest Colorado. It was laid down in the Early Permian during the Wolfcampian stage. Its subunits, therefore, are variously called formations or members depending on the publication. Members (youngest to oldest) include: De Chelly Sandstone (Arizona, Colorado, New Mexico, Utah); White Rim Sandstone (Utah); Organ Rock Shale (Arizona, Colorado, New Mexico, Utah); Cedar Mesa Sandstone (Arizona, Utah); Elephant Canyon Formation (Utah); Halgaito Shale (Arizona, Colorado, New Mexico, Utah (oldest).

White Rim Sandstone

The White Rim Sandstone is a sandstone geologic formation located in southeastern Utah. It is the last member of the Permian Cutler Group, and overlies the major Organ Rock Formation and Cedar Mesa Sandstone. It overlies thinner units of the Elephant Canyon and Halgaito Formations. The White Rim is eponymous, as the sandstone is named for its prominent white color and forms the rims of cliffs. It is the continental geologic formation deposited at the time of marine transgressions during the Early to Middle Permian Period. The White Rim Sandstone typically occurs above the Organ Rock Formation in southeast Utah which sits upon the extensive Cedar Mesa Sandstone in southeast Utah (**Figure 9**). Occurrences are in Moab, Utah about 25 miles (40 km) west of the Colorado border. The Circle Cliffs lie about 90 miles (145 km) southwest from Moab, and east of Escalante and Boulder, Utah (Utah State Route 12).

The Circle Cliffs are located in the northeast of Grand Staircase-Escalante National Monument. The cliffs also extend north-northwest adjacent the west perimeter of the Waterpocket Fold, and Capitol Reef National Park (**Figure 10**). The fold is traversed by Burr Trail-Straton Road which has views toward the Circle Cliffs, southward, or westward. The White Rim Road traverses the White Rim Sandstone Formation between the base of the Island in the Sky mesa and the Colorado and Green Rivers within Canyonlands National Park. In the Circle Cliffs (extreme south)-southeast-Utah the following formations occur: Moenkopi Formation; Organ Rock Formation; Cedar Mesa Sandstone; Moab, Utah — W. Colorado Moenkopi Formation.

Figure 9. White Rim Sandstone caps Organ Rock Shale in the Maze District of Canyonlands National Park. It is a close up view of Figure 8.

Figure 10. White Rim Sandstone exposures along the Burr Trail in the Waterpocket Fold in the Capital Reef National Park.

Organ Rock Shale

The Organ Rock Formation or Organ Rock Shale is a formation within the late Pennsylvanian to early Permian Cutler Group and is deposited across southeastern Utah, northwestern New Mexico, and northeastern Arizona. This formation notably outcrops around Canyonlands National Park, Natural Bridges National Monument, and Monument Valley of northeast Arizona, southern Utah. The age of the Organ Rock is constrained to the latter half of the Cisuralian epoch by age dates from overlying and underlying formations. Important early terrestrial vertebrate fossils have been recovered from this formation in northern Arizona, southern Utah, and northern New Mexico. These include the iconic Permian terrestrial fauna: *Seymouria*, *Diadectes*, *Ophiacodon*, and *Dimetrodon*. The fossil assemblage present suggests arid environmental conditions. This is corroborated with paleo-climate data indicative of global drying throughout the early Permian.

The Organ Rock Formation is present across southeastern Utah, U.S.A. It outcrops around Canyonlands National Park, Natural Bridges National Monument, and Monument Valley. In these areas the Organ Rock typically outcrops as a dark-red/brown siltstone to mudstone gently dipping towards the southeast. Within Canyonlands N.P. it forms towers which are meters to tens of meters tall. These are protected by caps of the White Rim Sandstone. In general, the Organ Rock Formation records the evolution of terminal fluvial fans which dry up into sections of overlying formations. Animals living during the time of the Organ Rock's deposition had the capacity to travel across most of earth's landmass as at the time land was all concentrated into the supercontinent Pangea. Figure 9 is a close up view of the Organ Rock Shale below the White rim Sandstone cap.

The Organ Rock Formation is conformably underlain by the Cedar Mesa Sandstone. It is conformably overlain by the De Chelly Sandstone around Monument Valley and by the White Rim Sandstone in the Canyonlands National Park (**Figure 11**). In locations where the De Chelly and White Rim are absent, the Organ Rock is unconformably overlain by the Triassic Moenkopi or Chinle Formations by an erosional contact. Toward its eastern extent, the Organ Rock Formation grades into the Cutler Formation, undivided. This transition occurs to the southwest of Moab, UT. The age of the Organ Rock Formation is unconfirmed. The preceding Cedar Mesa Sandstone is dated to the Wolfcampian Artinskian. The anteceding De Chelly and White Rim Sandstones are dated to the Leonardian (Kungurian). These formations constrain the age of the Organ Rock Formation to the latter part of the Cisuralian Epoch, approximately 290.1 to 272.3 Ma. The Organ Rock Formation may contain the Artinskian / Kungurian boundary. This is during the early to mid-Permian, a time where synapsids and temnospondyl amphibians were the dominant players in terrestrial ecosystems. These animals predate the advent of archosaur reptiles which give rise to dinosaurs in the Triassic.

Figure 11. The Organ Rock Shale is the sloping eroded talus slopes beneath the more resistant stands of De Chelly Sandstone buttes in Monument Valley. Organ Rock shale is darker red in color. De Chelly sandstone is the lighter orange colored units.

The Organ Rock Formation is composed of sandstones, siltstones, conglomerates, and mudstones. These rocks occur in two primary facies: floodplain and channel as well as eolian dunes and sand sheets. In these deposits mixed reddish-brown sands and silts dominate with minor mudstones and conglomerates interspersed. Interpreted channel deposits grade from laterally accreting coarse sands into conglomerates of pebble-sized carbonate clasts. Channels are 0.5 meter to 7 meters thick and may extend laterally for a few hundred meters. Fine-grained silt to mudstone deposits are proximal to interpreted channels. These deposits are interpreted as meandering streams with associated floodplains.

Eolian dune and sand sheet deposits are characterized by pale red fine- to medium-grained sandstones. These units are cross bedded. Each cross-bed is composed of thin, consistently spaced laminations. These strata are interpreted as those made by migrating dunes. This facies is marked by a sharp contact at the top of the preceding floodplain and channel facies. Mud cracks found at the top of the facies are filled by fine-grained sand from the overlying dune facies. This facies is most common at the western extent of the Organ Rock Formation.

Cedar Mesa Sandstone

Cedar Mesa Sandstone (also known as the Cedar Mesa Formation) is a sandstone member of the Cutler Formation found in southeast Utah, southwest Colorado, northwest New Mexico, and northeast Arizona. Cedar Mesa Sandstone are the remains of coastal sand dunes deposited about 245–286 million years ago, during the early Permian period. Coloration varies, but the rock often displays a red and white banded appearance as a result of periodic floods which carried iron-rich sediments down from the Uncompahgre Mountains during its formation. Named after topographic Cedar Mesa near the San Juan River in Utah, exposures of Cedar Mesa Sandstone form the spires and canyons found in the Needles and Maze districts of Canyonlands National Park, the inner gorge of White Canyon, and the three natural bridges of Natural Bridges National Monument (**Figure 12**).

Figure 12. The Cedar Mesa Formation caps the type section Cedar Mesa. It is a mixture of banded red and white sandstone. Source: Adriel Heisey posted on the internet.

Elephant Canyon Formation

The Elephant Canyon Formation is the basal Permian geologic formation of the Cutler Group overlying an unconformity on the Pennsylvanian Honaker Trail Formation in the Paradox Basin of southern Utah. **Figure 13** is a paleogeographic map showing the limits of the Elephant Canyon Formation in southeastern Utah.

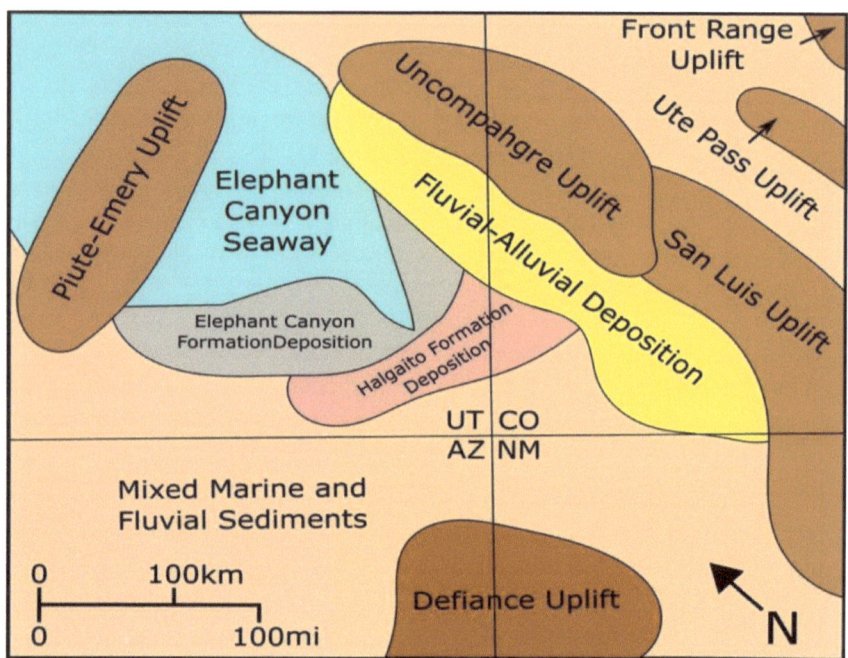

Figure 13. Paleogeographic map showing the limits of the Elephant Canyon formation in southeastern Utah. The Elephant Canyon Seaway was the source of deposition. The formation remnants are located at the southeastern edge of the seaway which were preserved in the landscape. The interior parts of the seaway were eroded since deposition. Source: BioOne posted on the internet.

Halgaito Shale

The Halgaito Shale is the basal Permian geologic member of the Cutler Group in southern Utah. The member consists of silty sandstone, siltstone and limestone. The Elephant Canyon may grade into the Halgaito and grades northward into the Cedar Mesa Formation. There is no designated type locality for the Halgaito. The shale can be seen at the confluence of the Green River and Colorado Rivers and in Cataract Canyon.

There is no designated type locality for the Cutler. It was named in 1905 after Cutler Creek which enters Uncompahgre River about 4 miles north of Ouray, Colorado (**Figure 14**). The formation was divided into the Halgaito Tongue (base), Cedar Mesa Sandstone Member, Organ Rock Tongue, and White Rim Sandstone Member. Cutler's geographic extent was established and raised the formation to group rank. In Arizona, Colorado, New Mexico and Utah, the Cutler Group occupies the Black Mesa Basin, Paradox Basin, Piceance Basin, San Juan Mountains province, San Juan Basin, and Uinta Basin.

Figure 14. Cutler Creek discharges into the Uncompahgre River in Colorado at the town of Portland, Colorado. The location exposes the Cutler Group as close to the "type" locality as possible although there is no type locality for the group.

Moenkopi Formation

Clastic red beds were laid down in shallow-water on top of the eroded Paleozoic surface early in the Triassic. These sediments were deposited on flood plains by streams as part of an expansive lowland that was slightly sloped in the direction of an ocean to the west. Mud built up in tidal flats became the mudstone of the Moenkopi Formation. Examples of this formation some of which still show fossilized ripple marks and mud cracks can be seen in the northern and western parts of the park (**Figure 15**).

The Moenkopi Formation is a geological formation that is spread across the U.S. states of New Mexico, northern Arizona, Nevada, southeastern California, eastern Utah and western Colorado. This unit is considered to be a group in Arizona. Part of the Colorado Plateau and Basin and Range, this red sandstone was laid down in the Lower Triassic and possibly part of the Middle Triassic around 240 million years ago.

There is no designated type locality for this formation. It was named for a development at the mouth of Moencopie Wash in the Grand Canyon area. A substitute type locality was located in the wall of the Little Colorado Canyon about 5 miles below Tanner Crossing in Coconino County, Arizona. While in the Great Basin, the formation was characterized and named the Rock Canyon Conglomerate, Virgin Limestone, and Shnabkaib Shale members.

Figure 15. Moenkopi Formation (red beds) were eroded before the overlying Pleistocene conglomerate bed was deposited above in central Utah. Source: Marli Miller posted on the internet.

Salt Creek (later replaced by Wupatki and Moqui Members) and the Holbrook Member were found and named in the Black Mesa basin. The Sinbad Limestone Member was named in the Paradox Basin. The Timpoweap Member in the Plateau sedimentary province, and the Wupatki Member was first used in the Plateau sedimentary province and its age was modified to Early and Middle (?) Triassic.

The Tenderfoot, Ali Baba, Sewemup, and Pariott Members were named in the Piceance and Uinta Basins. The Hoskinnini Member was assigned in the Black Mesa and Paradox basins. The Black Dragon, Torrey, and Moody Canyon members in the Paradox Basin and Plateau sedimentary province were named but were revised in 1979. The age was modified to Early and Middle Triassic using biostratigraphic dating in 1988. The Anton Chico Member was assigned in the Palo Duro Basin.

The Moenkopi consists of thinly bedded sandstone, mudstone, and shale, with some limestone in the Capitol Reef area. It has a characteristic deep red color and tends to form slopes and benches. The depositional environment varies from fluvial channel and flood plain deposits in the eastern exposures to tidal mudflats in the Cedar Mesa area to deltaic sandstones and shallow marine limestones at Capitol Reef. In eastern Nevada and northwestern Utah, it thickens dramatically then transitions to the Woodside, Thaynes, and Mahogony formations. The general deposition setting was sluggish rivers traversing a flat featureless coastal plane to the sea.

The low relief meant that the shoreline moved great distances with changes of sea level or even with the tides. Thickness varies from a feather edge against the Uncompahgre highlands to the east to over 2,000 feet (610 meters) in southwestern Utah. The thickness varies greatly in the Paradox Basin where the Moenkopi is thin to nonexistent on the crests of salt anticlines and over 400 meters (1,300 feet) thick in the corresponding synclines.

The Moenkopi rests unconformably on Paleozoic beds and the Chinle Formation in turn rests unconformably on the Moenkopi. Both unconformities are locally angular unconformities. The lower unconformity corresponds to the regional Triassic 1 unconformity and the upper to the regional Triassic 3 unconformity. The Triassic 1 unconformity represents a hiatus of at least 20 million years while Triassic 2 represents a hiatus of about 10 million years.

Members differ considerably from east to west in part because sandstone beds corresponding to marine transgressions are used to define members to the west but cannot be traced to the east. In different regions by ascending stratigraphic order the members are designated as: Paradox Basin: Tenderfoot Member. This is everywhere in contact with the Cutler Formation on an angular unconformity. Up to 290 feet (88 meters) thick, it consists of basal conglomerate, gypsum, massive cliff-forming mudstone and silty sandstone.

Ali Baba Member. This is red conglomeratic sandstone and red siltstone and was laid down by energetic rivers. A distinctive feature is load structures. Sediment sources were apparently the crests of anticlines, and paleo-current directions are north to northwest along the corresponding synclines.

Sewemup Member. This is thinly bedded siltstone, shale, and sandstone with enough gypsum content to give it a light brown color that contrasts with the darker brown Ali Baba Member.

Parriott Member. This is distinguished from the light brown Sewemup Member by its multihued brown, red, orange and purple strata. Exposures are geographically limited appearing only in Richardson Amphitheater, Castle Valley, Sinbad Valley, and Big Bend meander of the Colorado River. This member may be separated from the Sewemup Member by a regional unconformity.

In the Canyonlands and Glen Canyon area, the Moenkopi Formation is divided into the *Hoskinnini Member*: Sandstone and siltstone. Found only in the southern part of the area.

Black Dragon Member. Consists of a basal conglomerate; thinly bedded red sandstone, siltstone, and shale deposited in a tidal flat environment; a sandstone sheet; and a second sequence of tidal flat deposits.

Sinbad Limestone Member. Named for the Sinbad region in the San Rafael Swell. Consists of yellowish limestone deposited by a brief marine transgression.

Torrey Member. Red beds signifying a return to subaerial deposition.

Moody Canyon Member. Thinly bedded slope forming siltstone and mudstone with minor evaporites.

In the San Juan Basin and Tucumcari area, the Moenkopi Formation is divided into the following members:

Anton Chico Member. Described previously as the red sandstone member of the Santa Rosa Formation, but placed in its own formation after it was recognized to be middle Triassic in age. Subsequently named as a member of the Moenkopi when its beds were traced west and demonstrated that it was deposited in a single basin with Moenkopi beds in Arizona. Other members listed in alphabetical order with asterisks (*) indicating usage by the U.S. Geological Survey and other usages by state geological surveys:

Holbrook Sandstone Member (AZ*); *Moqui Member* (AZ*); *Rock Canyon Conglomerate Member* (AZ*, NV*, UT*); *Shnabkaib Member* (AZ*, NV*, UT*); *Timpoweap Member* (AZ, NV, UT*); *Virgin Limestone Member* (AZ*, NV*, UT*); and *Winslow Member* (AZ).

The Moenkopi formation is found in these geologic locations: Black Mesa Basin*; Great Basin province*; Green River Basin*; Las Vegas-Raton Basin; Orogrande Basin; Palo Duro Basin; Paradox Basin*; Piceance Basin*; Plateau sedimentary province*; San Juan Basin*; Uinta Basin*. The asterick refers to USGS designates.

The Moenkopi Formation is found within these parks (incomplete list): Grand Canyon National Park; Capitol Reef National Park; Zion National Park; Monument Valley Navajo Nation Tribal Park; Dinosaur National Monument; Glen Canyon National Recreation Area; Walnut Canyon National Monument; and Wupatki National Monument.

Chinle Formation

Another period of erosion returned creating a third unconformity. The brightly colored shales of the slope-forming Chinle Formation were laid down on top of this eroded surface. Petrified wood from the Petrified Forest Member of the Chinle is sometimes found at the base of Chinle slopes. The Chinle Formation was deposited in the Lower Triassic Period (**Figure 16**).

Figure 16. The Chinle Formation belongs to the slope forming deposits at the base of the cliff face. Source: Ames Geology-Weebly posted on the internet.

The Chinle Formation is an Upper Triassic continental geological formation of fluvial, lacustrine, and palustrine to eolian deposits spread across the U.S. states of Nevada, Utah, northern Arizona, western New Mexico, and western Colorado. The Chinle is controversially considered to be synonymous to the Dockum Group of eastern Colorado and New Mexico, western Texas, the Oklahoma panhandle, and southwestern Kansas. The Chinle is sometimes colloquially named as a formation within the Dockum Group in New Mexico and in Texas. The Chinle Formation is part of the Colorado Plateau, Basin and Range, and the southern section of the Interior Plains. A probable separate depositional basin within the Chinle is found in northwestern Colorado and northeastern Utah. The southern portion of the Chinle reaches a maximum thickness of a little over 520 m. Typically, the Chinle rests unconformably on the Moenkopi Formation.

There is no designated type localities for this formation. It was named for Chinle Valley in Apache County, Arizona without officially designating it as the formation's name and without a specified type locality. The United States Geological Survey revised the unnamed members by identifying and naming the Temple Mountain member as the basal-most unit in the area of the San Rafael Swell of Utah. The Shinarump Conglomerate was renamed Shinarump member of the Chinle formation.

The formation members and their thicknesses are highly variable across the Chinle. The stratigraphically lowest formation is the Temple Mountain Member. However, in most areas, the basal member is the Shinarump Member. The Shinarump is a braided-river system channel deposit facies. The Monitor Butte Member overlies the Shinarump in most areas. The Monitor Butte is an overbank (distal floodplain) facies with lacustrine deposits. This is overlain in western areas by the channel-deposit facies Moss Back Member. More commonly, the Monitor Butte grades into the Petrified Forest Member. The Petrified Forest is predominately over bank deposits with thin lenses of channel-deposit facies and lacustrine deposits.

The Petrified Forest Member can be divided into the Lower Petrified Forest Member and Upper Petrified Forest Member in Arizona and New Mexico separated by the Sonsela Sandstone. The Sonsela Sandstone is a braided-stream channel facies. The Petrified Forest grades into the Owl Rock Member. The Owl Rock is a marginal lacustrine to lacustrine facies possibly representing a large lake system. The upper Petrified Forest Member includes the thin but extensive Correo Sandstone Bed.

Finally, either the Rock Point or Church Rock Members overlie the Owl Rock. Some researchers feel the two Members may be synonymous. The two Members are complex heterolithic units representing variously braided-river facies, lacustrine, and over bank deposits. In 2020, the names Blue Mesa Member and Painted Desert Member were proposed for the lower and upper Petrified Forest members in west-central New Mexico where the Chinle Formation was raised to group rank. Beds between the Shinarump Formation and the Petrified Forest Formation were assigned to the Bluewater Creek Formation. The Chinle Formation is aged from early Late Triassic. Age correlation is based on Land Vertebrate Fauna-chrons. The fauna-chrons are based on first and last appearances of phytosaurs.

Asterisks (*) indicate usage by the U.S. Geological Survey. Other usages by state geological surveys may occur. Group rank (alphabetical – rank and formations not recognized by the USGS): San Pedro Arroyo Formation (NM); Santa Rosa Formation (NM); Shinarump Formation (NM); Petrified Forest Formation (AZ, UT, NM).

Many others were assigned to the Chinle formation but several members are not recognized by the USGS: Agua Zarca Sandstone Member (NM); Bluewater Creek Member (AZ, CO, NM); Church Rock Member (AZ*, CO*, UT*); Correo Sandstone Member (NM); Cuervo Sandstone Member (NM); Duffin Sandstone Member (UT); Gartra Member (CO*, UT*); Mesa Redondo Member (AZ*, NM*); Monitor Butte Member (AZ*, CO*, UT*); Moss Back Member (AZ*, CO*, UT*); Newspaper Rock Sandstone Bed (AZ); Owl Rock Member (AZ*, NM*, UT*); Petrified Forest Member (AZ*, CO*, NV*, NM*, UT*); Poleo Sandstone Lentil (NM); Redonda Member (NM); Rock Point Member (AZ*, NM*); Salitral Shale Tongue (NM*); Shinarump Member (AZ*, NV*, NM*, UT*); Silver Reef Sandstone Member (UT); Stanaker Member (UT); Temple Mountain Member (UT*); Trail Hill Sandstone [Member] (UT).

Geologic Provinces where the Chinle formation occurs: Black Mesa Basin*; Great Basin province*; Green River Basin*; Las Vegas-Raton Basin*; Orogrande Basin; Palo Duro Basin*; Paradox Basin*; Permian Basin; Piceance Basin*; Plateau sedimentary province*; San Juan Basin*; Sierra Grande Uplift*; Uinta Basin*; Uinta Uplift*; Wasatch Uplift*.

Parklands where the Chinle formation occurs: Arches National Park; Canyonlands National Park; Capitol Reef National Park; Colorado National Monument; Dinosaur National Monument; Flaming Gorge National Recreation Area; Glen Canyon National Recreation Area; Gold Butte National Monument; Grand Canyon National Park; Grand Staircase-Escalante National Monument; Lake Mead National Recreation Area; Natural Bridges National Monument; Petrified Forest National Park; Red Fleet State Park; Wupatki National Monument; Zion National Park and the Kolob canyons area; and, Red Rock Canyon National Conservation Area.

The Glen Canyon Group

The Glen Canyon Group of formations includes (from oldest: lowest to youngest): the Wingate Sandstone, Kayenta Formation, and the Navajo Sandstone. These formations are most prominently exposed in the western and northern sections of the park. Triassic climates progressively became dryer prompting the formation of sand dunes that buried dry stream beds and their flood plain. This sand became the cliff-forming deposit of several hundred feet (many tens of meters) high and red-colored Windgate Sandstone. Outcrops tend to run for hundreds of miles (hundreds of kilometers) with few breaks creating an impediment to human travel (**Figure 17**).

For a time climatic conditions became wetter and streams cut channels through the sand dunes. Reddish-brown to lavender-colored sandstones interbedded with siltstones and shales constitute the resulting ledgy slope forming Kayenta Formation. The youngest and therefore topmost formation in the Glen Canyon Group was formed after arid conditions returned to the region. A vast and very dry desert, not unlike the modern Sahara, covered 150,000 square miles (388,000 km^2) of western North America.

Figure 17. The Glen Canyon Group is represented by the white to tan Navajo Formation (N) in the middle, the Wingate Sandstone in the central wedge shaped deposit (W), and the Kayenta Formation on top forming the cliffs (K). Location is along Burr Trail in between Capita Reef National park and Glen Canyon National Recreation Area. Source: Written in Stone As Seen Through My Lens blog posted on the internet.

Cross bedded sand dunes accumulated to great thickness especially in the nearby Zion and Kolob canyons area forming the buff to pale orange Navajo Sandstone. Navajo outcrops form cliffs, temples, and under certain conditions natural arches (such as Millard Canyon Arch) in the area.

A fourth unconformity was created by a period of erosion. Mud flats developed on top of the eroded surface of the Glen Canyon Group forming the Carmel Formation. The massive cliff-forming Entrada Sandstone in turn was created on top of the Carmel. A long period of erosion stripped away most of the San Rafael Group in the area along with any formations that may have been laid down in the Cretaceous period (**Figure 18**).

The Wingate Formation

The Wingate Sandstone is a geologic formation in the Glen Canyon Group of the Colorado Plateau province of the United States which crops out in northern Arizona, northwest Colorado, Nevada, and Utah. Wingate Sandstone is particularly prominent in southeastern Utah where it forms attractions in a number of national parks and monuments. These include Capitol Reef National Park, the San Rafael Swell, and Canyonlands National Park.

Figure 18. The Carmel Formation is exposed in this road cut along Interstate 70 in the San Rafael Swell in southeastern Utah. Source: Elizabeth Petrie posted on the internet.

Wingate Sandstone frequently appears just below the Kayenta Formation and Navajo Sandstone, two other formations of the Glen Canyon group. Together, these three formations can result in immense vertical cliffs of 2000 feet (609 meters) or more. Wingate layers are typically pale orange to red in color, the remnants of wind-born sand dunes deposited approximately 200 million years ago in the Late Triassic.

The Kayenta Formation

The Kayenta Formation is a geologic layer in the Glen Canyon Group that is spread across the Colorado Plateau province of the United States including northern Arizona, northwest Colorado, Nevada, and Utah. This rock formation is particularly prominent in southeastern Utah where it is seen in the main attractions of a number of national parks and monuments. These include Zion National Park, Capitol Reef National Park, the San Rafael Swell, and Canyonlands National Park.

The Kayenta Formation frequently appears as a thinner dark broken layer below Navajo Sandstone and above Wingate Sandstone (all three formations are in the same group). Together, these three formations can result in immense vertical cliffs of 2,000 feet (610 m) or more. Kayenta layers are typically red to brown in color, forming broken ledges.

In most sections in southeastern Utah that include all three geologic formations of the Glen Canyon Group, the Kayenta is easily recognized. Even at a distance it appears as a dark-red, maroon, or lavender band of thin-bedded material between two thick, massive cross bedded strata of buff, tan, or light-red color. Its position is also generally marked by a topographic break. Its weak beds form a bench or platform developed by stripping the Navajo sandstone back from the face of the Wingate cliffs. The Kayenta is made up of beds of sandstone, shale, and limestone, all lenticular, uneven at their tops, and discontinuous within short distances. They suggest deposits made by shifting streams of fluctuating volume. The sandstone beds from less than 1-inch (25 mm) to more than 10 feet (3.0 m) thick are composed of relatively coarse well-rounded quartz grains cemented by lime and iron. The thicker beds are indefinitely cross bedded. The shales are essentially fine-grained very thin sandstones that include lime concretions and balls of consolidated mud. The limestone appears as solid gray-blue beds a few inches to a few feet thick, and as lenses of limestone conglomerate. Most of the limestone lenses are less than 25 feet (7.6 m) long, but two were traced for nearly 500 feet (150 m) and one for 1,650 feet (500 m).

Viewed as a whole, the Kayenta is readily distinguished from the geologic formations above and below it. It is unlike them in composition, color, manner of bedding, and sedimentary history. Obviously the conditions of sedimentation changed in passing from the Wingate Sandstone formation to the Kayenta and from the Kayenta to the Navajo sandstone, but the nature and regional significance of the changes have not been determined. In some measured sections the transition from Wingate to Kayenta is gradual. The material in the basal Kayenta, beds seems to have been derived from the Wingate immediately below and redeposited with only the discordance characteristic of fluviatile sediments. But in many sections the contact between the two formations is unconformable.

The basal Kayenta consists of conglomerate and lenticular sandstone that fills depressions eroded in the underlying beds. In Moqui Canyon near Red Cone Spring nearly 10 feet (3.0 m) of Kayenta limestone conglomerate rests in a long meandering valley cut in Wingate. Likewise, the contact between the Kayenta and the Navajo in places seems to be gradational, but generally a thin jumbled mass of sandstone and shales, chunks of shale and limestone, mud balls, and concretions of lime and iron lies at the base of the fine-grained cross bedded Navajo. Mud cracks, a few ripple marks, and incipient drainage channels were observed in the topmost bed of the Kayenta on Red Rock Plateau. In west Glen Canyon, wide sand-filled cracks appear at the horizon. These features indicate that in places at least the Wingate and Kayenta were exposed to erosion before their overlying geologic formations were deposited, or it may be that the range in thickness of the Kayenta thus in part (is) accounted for.

In southwestern Utah, the red and mauve Kayenta siltstones and sandstones that form the slopes at base of the Navajo Sandstone cliffs record the record of low to moderate 210 m (500 to about 700 feet) of sediment here. The sedimentary structures showing the channel and flood plain deposits of streams are well exposed on switchbacks below the tunnel in Pine Creek Canyon (**Figure 19**).

Figure 19. The red and mauve cliffs belong to the Kayenta Formation in southwestern Utah. The location is in Pine Valley east of Central Utah in the Dixie National Forest.

In the southeastern part of Zion National Park, a stratum of cross bedded sandstone is found roughly halfway between the top and bottom of the Kayenta Formation. It is a "tongue" of sandstone that merges with the Navajo formation east of Kanab, and it shows that desert conditions occurred briefly in this area during Kayenta time. This tongue is the ledge that shades the lower portion of the Emerald Pool Trail, and it is properly called Navajo, not Kayenta.

Fossil mud cracks attest to occasional seasonal climate, and thin limestones and fossilized trails of aquatic snails or worms mark the existence of ponds and lakes. The most interesting fossils, however, are the dinosaur tracks that are relatively common in Kayenta mudstone. These vary in size, but all seem to be the tracks of three-toed reptiles that walked upright, leaving their tracks in the muds on the flood plains. Unfortunately, so far no bone materials have been found in Washington County that would enable more specific identification.

Apparently during Kayenta time, Zion was situated in a climatic belt like that of Senegal with rainy summers and dry winters at the southern edge of a great desert. The influence of the desert was about to predominate, however, as North America drifted northward into the arid desert belt.

In the Glen Canyon, the Kayenta Formation is approximately 400 feet (120 m) thick and consists of a fine grained sandstone interbedded with layers of siltstone. The alternation of these units generally produces a series of ledges and slopes between the cliffs of the Navajo and Moenave formation. Dinosaur tracks are fairly common in the siltstone, and fresh water mussels and snails occur but are rare. The Kayenta Formation is colored pale red and adds to the splendor of the Vermilion Cliffs. It accumulated as deposits of rivers.

Navajo Sandstone

The Navajo Sandstone is a geological formation in the Glen Canyon Group that is spread across the U.S. states of southern Nevada, northern Arizona, northwest Colorado, and Utah as part of the Colorado Plateau province of the United States. The Navajo Sandstone is particularly prominent in southern Utah where it forms the main attractions of a number of national parks and monuments including Red Rock Canyon National Conservation Area, Zion National Park, Capitol Reef National Park, Glen Canyon National Recreation Area, Grand Staircase-Escalante National Monument, and Canyonlands National Park. Navajo Sandstone frequently overlies and inter-fingers with the Kayenta Formation of the Glen Canyon Group. Together, these formations can result in immense vertical cliffs of up to 2,200 feet (670 m). Atop the cliffs, Navajo Sandstone often appears as massive rounded domes and bluffs that are generally white in color.

Navajo Sandstone frequently occurs as spectacular cliffs, cuestas, domes, and bluffs rising from the desert floor. It can be distinguished from adjacent Jurassic sandstones by its white to light pink color, meter-scale cross-bedding, and distinctive rounded weathering. The wide range of colors exhibited by the Navajo Sandstone reflect a long history of alteration by groundwater and other subsurface fluids over the last 190 million years. The different colors, except for white, are caused by the presence of varying mixtures and amounts of hematite, goethite, and limonite filling the pore space within the quartz sand comprising the Navajo Sandstone. The iron in these strata originally arrived via the erosion of iron-bearing silicate minerals. Initially, this iron accumulated as iron-oxide coatings which formed slowly after the sand had been deposited. Later, after having been deeply buried, reducing fluids composed of water and hydrocarbons flowed through the thick red sand which once comprised the Navajo Sandstone. The dissolution of the iron coatings by the reducing fluids bleached large volumes of the Navajo Sandstone a brilliant white. Reducing fluids transported the iron in solution until they mixed with oxidizing groundwater.

Where the oxidizing and reducing fluids mixed, the iron precipitated within the Navajo Sandstone. Depending on local variations within the permeability, porosity, fracturing, and other inherent rock properties of the sandstone, varying mixtures of hematite, goethite, and limonite precipitated within spaces between quartz grains. Variations in the type and proportions of precipitated iron oxides resulted in the different black, brown, crimson, vermillion, orange, salmon, peach, pink, gold, and yellow colors of the Navajo Sandstone. The precipitation of iron oxides also formed lamina, corrugated layers, columns, and pipes of ironstone within the Navajo Sandstone. Being harder and more resistant to erosion than the surrounding sandstone, the ironstone weathered out as ledges, walls, fins, "flags", towers, and other minor features which stick out and above the local landscape in unusual shapes.

The age of the Navajo Sandstone is somewhat controversial. It may have originated from the Late Triassic but is at least as young as the Early Jurassic stages Pliensbachian and Toarcian. There is no type locality of the name. It was simply named for the 'Navajo Country' of the southwestern United States. The two major subunits of the Navajo are the Lamb Point Tongue (Kanab area) and the Shurtz Sandstone Tongue (Cedar City area) (**Figure 20**).

Figure 20. The Shurtz Sandstone tongue near Cedar City on Center Road is exposed and is part of the Navajo Sandstone formation.

The Navajo Sandstone was originally named as the uppermost formation of the La Plata Group but later was reassigned as the upper formation of Glen Canyon Group. The name was originally not used in northwest Colorado and northeast Utah where the name 'Glen Canyon Sandstone' was preferred. Radioisotopic analysis suggests that the Navajo Sandstone formation is entirely Jurassic extending for about 5.5 million years from the Hettangian age to the Sinemurian age. The sandstone was deposited in an arid erg on the Western portion of the Supercontinent Pangaea. This region was affected by annual monsoons that came about each winter when cooler winds and wind reversal occurred. Navajo Sandstone outcrops are found in these geologic locations: Colorado Plateau; Black Mesa Basin; Great Basin province; Paradox Basin; Piceance Basin; Plateau sedimentary province; San Juan Basin; Uinta Basin; Uinta Uplift; Uncompaghre Uplift;

The formation is also found in these parklands (incomplete list): Glen Canyon National Recreation Area; Grand Staircase-Escalante National Monument; Zion National Park; Canyonlands National Park; Capitol Reef National Park; Arches National Park; Dinosaur National Monument; Navajo National Monument; Colorado National Monument; and, Pink coral sand dunes, Kanab, Utah.

Iron Oxide Concretions

The Navajo Sandstone is also well known among rock hounds for its hundreds of thousands of iron oxide concretions. Informally, they are called "Moqui marbles" and are believed to represent an extension of Hopi Native American traditions regarding ancestor worship ("moqui" translates to "the dead" in the Hopi language) (**Figure 21**). Thousands of these concretions weather out of outcrops of the Navajo Sandstone within south-central and southeastern Utah within an area extending from Zion National Park eastward to Arches and Canyonland national parks. They are quite abundant within Grand Staircase-Escalante National Monument.

The iron oxide concretions found in the Navajo Sandstone exhibit a wide variety of sizes and shapes. Their shape ranges from spheres to discs, buttons, spiked balls, cylindrical hollow pipe-like forms, and other odd shapes. Although many of these concretions are fused together like soap bubbles, many more also occur as isolated concretions which range in diameter from the size of peas to baseballs. The surface of these spherical concretions can range from being very rough to quite smooth. Some of the concretions are grooved spheres with ridges around their circumference.

The abundant concretions found in the Navajo Sandstone consist of sandstone cemented together by hematite (Fe_2O_3), and goethite ($FeOOH$). The iron forming these concretions came from the breakdown of iron-bearing silicate minerals by weathering to form iron oxide coatings on other grains. During later diagenesis of the Navajo Sandstone while deeply buried, reducing fluids likely hydrocarbons dissolved these coatings. When the reducing fluids containing dissolved iron mixed with oxidizing groundwater both were oxidized.

Figure 21. Iron oxide concretions called Moqui marbles eroded from Navajo Sandstone were deposited while deposition of the sandstone occurred. Ground water and surface water appear to play a role but the pattern of the concretions mimics polygonal shapes possibly associated with fractures and other pathways through the sand deposits. Monsoon rainfall may have infiltrated these pathways thus concentrating iron and other dissolved minerals along these pathways to form concretions. Source: Posted on the internet.

This caused the iron to precipitate out as hematite and goethite to form the innumerable concretions found in the Navajo Sandstone. Evidence suggests that microbial metabolism may have contributed to the formation of some of these concretions. These concretions are regarded as terrestrial analogues of the hematite spherules called alternately Martian "blueberries" or more technically Martian spherules which the Opportunity rover found at Meridiani Planum on Mars.

The Entrada Sandstone

The Entrada Sandstone is a formation in the San Rafael Group found in the U.S. states of Wyoming, Colorado, northwest New Mexico, northeast Arizona, and southeast Utah. Part of the Colorado Plateau, this formation was deposited during the Jurassic period sometime between 180 and 140 million years ago in various environments including tidal mudflats, beaches, and sand dunes. The Middle Jurassic San Rafael Group was dominantly deposited as ergs (sand seas) in a desert environment around the shallow Sundance Sea.

This formation has been dated to the early to middle Callovian stage of the latest Middle Jurassic. The type locality and place for which the unit is named is Entrada Point located in the northern part the San Rafael Swell in Emery County, Utah. The Entrada Sandstone was named as one of the four formations of the San Rafael Group. In the original description, the Entrada is overlain by the Curtis Formation and overlies the Carmel Formation. In the Curtis Mountains region of northeastern Arizona, the Entrada is overlain by the Wanakah Formation. The geographic extent of the formation was revised several times afterwards. It was divided into the Gunsight Butte, Cannonville, and Escalante members (**Figure 22**).

Figure 22. The Entrada Formation occurs between the lower Carmel formation and upper Curtis Formation in Arches National Park. An unconformity surface separates the Entrada from the Carmel, and the Entrada from the overlying Curtis Formation. Source: National Park Service posted on the internet.

Entrada members are (in alphabetical order): Cannonville Member (UT); Cow Springs Member (AZ); Dewey Bridge Member (CO, UT) - named after the type locality at Dewey Bridge. This brick-red layer has a blocky look to it; Escalante Member (UT); Henrieville Member (UT); Exeter Member (NM); Gunsight Butte Member (UT); Iyanbito Member (NM); Moab Member (CO, UT) or Moab Tongue (CO, UT) – named after the type locality of Moab, Utah. The whitish sands from inland dunes make up this "cap rock" layer as seen atop Delicate Arch and Broken Arch in Arches National Park; Red Mesa Member (AZ, NM, UT); Slick Rock Member (CO, UT) - named for the type locality at Slick Rock, Colorado; rounded beach sands were cemented together to create this uniform layer.

Entrada Sandstone is found in these geologic provinces: Anadarko Basin; Black Mesa Basin; Denver Basin; Great Basin province; Green River Basin; Las Vegas-Raton Basin; Paradox Basin; Piceance Basin; Plateau sedimentary province; San Juan Basin; Sierra Grande Uplift. Entrada Sandstone is found in these parklands (incomplete list): Arches National Park; Capitol Reef National Park; Goblin Valley State Park; and, Kodachrome Basin State Park (**Figure 23**).

Figure 23. Kodachrome Basin State Park, Utah is located about 20 miles southeast of Bryce Canyon National Park in Cannonville Utah. The White Cliffs are in the background and the northwestern edge of the Red Cliffs are in the middle and foreground.

Chapter 3. Uplifts

Starting 70 million years ago and extending well into the Cenozoic, a mountain-building event called the Laramide orogeny uplifted the Rocky Mountains and with it the Canyonlands region. Even though the strata were uplifted thousands of feet (hundreds of meters) they were left at more-or-less the same horizontally. Uplift-associated-jointing did occur and has since influenced erosional patterns. When ground water seeped into the salt beds of the Paradox Formation it carried away the topmost and more soluble salts, leaving gypsum. This process was so pronounced in "The Grabens" that the overlying rock collapsed into voids left by escaping salt (**Figure 24**).

Figure 24. The Grabens in Canyonlands National Park formed by flowing salts beneath the more resistant rock formation above that led to the collapse of the rock formations where the flowing salt vacated the subsurface.

Increased precipitation during the ice ages of the Pleistocene quickened the rate of canyon excavation. Canyon widening and deepening was especially rapid for the gorges of the Green and Colorado Rivers which were in part fed by glacier-melt from the Rocky Mountains. Alluvial-fan-creation-landslides and sand dune-migration were also accelerated in the Pleistocene. These processes continue to shape the Canyonlands landscape in the Holocene (the current epoch), but at a slower rate due to a significant increase in aridity.

The Laramide Orogeny

The Laramide orogeny was a time period of mountain building in western North America which started in the Late Cretaceous 70 to 80 million years ago, and ended 35 to 55 million years ago. The exact duration and ages of beginning and end of the orogeny are in dispute. The Laramide orogeny occurred in a series of pulses with quiescent phases intervening. The major feature that was created by this orogeny was deep-seated, thick-skinned deformation with evidence of this orogeny found from Canada to northern Mexico. The easternmost extent of the mountain-building is represented by the Black Hills of South Dakota. The phenomenon is named for the Laramie Mountains of eastern Wyoming. The Laramide orogeny is sometimes confused with the Sevier orogeny which partially overlapped in time and space.

The orogeny is commonly attributed to events off the west coast of North America when the Kula and Farallon Plates were sliding under the North American plate. Most hypotheses propose that oceanic crust was undergoing flat-slab subduction, i.e., with a shallow subduction angle, and as a consequence no magmatism occurred in the central west of the continent. The underlying oceanic lithosphere actually caused drag on the root of the overlying continental lithosphere. One cause for shallow subduction may have been an increased rate of plate convergence.

Another proposed cause was subduction of thickened oceanic crust. Magmatism associated with subduction occurred not near the plate edges (as in the volcanic arc of the Andes, for example), but far to the east called the Coast Range Arc. Geologists call such a lack of volcanic activity near a subduction zone a *magmatic gap*. This particular gap may have occurred because the subducted slab was in contact with relatively cool continental lithosphere not hotter asthenosphere. One result of shallow angle of subduction and the drag that it caused was a broad belt of mountains some of which were the progenitors of the Rocky Mountains. Part of the proto-Rocky Mountains would be later modified by extension to become the Basin and Range Province.

The Laramide orogeny produced intermontane structural basins and adjacent mountain blocks by means of deformation. This style of deformation is typical of continental plates adjacent to convergent margins of long duration that have not sustained continent/continent collisions. This tectonic setting produces a pattern of compressive uplifts and basins with most of the deformation confined to block edges. Twelve kilometers of structural relief between basins and adjacent uplifts is not uncommon. The basins contain several thousand meters of Paleozoic and Mesozoic sedimentary rocks that predate the Laramide orogeny. As much as 5,000 meters (16,000 ft) of Cretaceous and Cenozoic sediments filled these orogenically-defined basins. Deformed Paleocene and Eocene deposits record continuing orogenic activity.

During the Laramide orogeny, basin floors and mountain summits were much closer to sea level than today. After the seas retreated from the Rocky Mountain region, floodplains, swamps, and vast lakes developed in the basins. Drainage systems imposed at that time persist today. Since the Oligocene, episodic epeirogenic uplift gradually raised the entire region including the Great Plains to present elevations. Most of the modern topography is the result of Pliocene and Pleistocene events including additional uplift, glaciation of the high country, and denudation and dissection of older Cenozoic surfaces in the basin by fluvial processes.

In the United States, these distinctive intermontane basins occur principally in the central Rocky Mountains from Colorado and Utah (Uinta Basin) to Montana and are best developed in Wyoming with the Bighorn, Powder River, and Wind River being the largest. Topographically, the basin floors resemble the surface of the western Great Plains except for vistas of surrounding mountains.

At most boundaries Paleozoic through Paleogene units dip steeply into the basins off uplifted blocks cored by Precambrian rocks. The eroded steeply dipping units form hogbacks and flatirons. Many of the boundaries are thrust or reverse faults. Although other boundaries appear to be monoclinal flexures, faulting is suspected at depth. Most bounding faults show evidence of at least two episodes of Laramide (Late Cretaceous and Eocene) movement suggesting both thrust and strike-slip types of displacement.

Chapter 4. Canyonlands National Park Field Tour

Geographic Setting

Lohamn (1974) provided a fairly complete geologic story of Canyonlands National Park before the park was improved with paved roads to various sites. The descriptions are provided in this summary with the caution that routes described may have changed since the publication was released.

Geologists divided the United States into many provinces each of which has distinctive geologic and topographic characteristics that set it apart from the others. One of the most intriguing and scenic of these is the Colorado Plateaus province referred to in this report simply as the Colorado Plateau, or the Plateau. This province which covers some 150,000 square miles and is not all plateaus as shall be seen extends from Rifle, Colo. at the northeast to a little beyond Flagstaff, Ariz. at the southwest, and from Cedar City, Utah at the west, nearly to Albuquerque, N. Mex., at the southeast. Canyonlands National Park appropriately occupies the heart of the Canyon Lands section, one of the six subdivisions of the Plateau. As the names implies, the Canyon Lands section of the Plateau comprises a high plateau generally ranging in altitude from 5,000 to 7,000 feet, which has been intricately dissected by literally thousands of canyons. Canyonlands National Park is drained entirely by the Colorado and Green Rivers whose confluence is an important and scenic central feature of the park. Individual canyons traversed or drained by these rivers are discussed in later chapters.

When Major Powell reached the confluence in 1869, the river flowing in from the northeast to join the Green River was called the Grand River, and the Green and Grand joined there to form the Colorado River. The Grand River was renamed Colorado River by act of the Colorado State Legislature approved March 24, 1921, and by act of Congress approved July 25, 1921. But the old term still remains in names such as Grand County, Colo., the headwaters region; Grand Valley, a town 16 miles west of Rifle, Colo.; Grand Valley between Palisade and Mack, Colo.; Grand Mesa which towers more than a mile above the Grand and Gunnison River valleys; Grand Junction, Colo., a city appropriately located at the confluence of the Grand and Gunnison Rivers; Grand County, Utah which the river traverses after entering Utah; and Grand View Point, the southern terminus of Island in the Sky.

Rocks and Landforms

The vivid and varied colors of the bare rocks and the fantastic canyons, buttes, spires, columns, alcoves, caves, arches, and other erosional forms of the canyon country result from a fortuitous combination of geologic and climatic circumstances and events unequaled in most other parts of the world. First among these events was the piling up, layer upon layer of thousands of feet of sedimentary rocks under a wide variety of environments.

Sedimentary rocks of the region are composed of particles ranging in size from clay and silt through sand and gravel carried to their resting places by moving water, silt and sand particles transported by wind, and some materials precipitated from water solutions such as limestone (calcium carbonate), dolomite (calcium and magnesium carbonate), gypsum (calcium sulfate with some water), anhydrite (calcium sulfate alone), common salt (sodium chloride), potash minerals such as potassium chloride, and a few other less common types. Some of the materials were laid down in shallow seas that once covered the area or in lagoons and estuaries near the sea. Some beds were deposited by streams in inland basins or plains. A few were deposited in lakes, and some like the Navajo Sandstone were carried in by the desert winds.

Not exposed in the area but present far beneath the sedimentary cover, and exposed in a few surrounding places are examples of the other two principal types of rocks: (1) igneous rocks, solidified from molten rock forced into or above younger rocks along cracks, joints, and faults, and (2) much older metamorphic rocks formed from other pre-existing rock types by great heat and pressure at extreme depths. The particles comprising the sedimentary rocks were derived by weathering and erosion of rocks of all three types in the headwater regions of the ancestral Colorado River basin. Igneous rocks of Tertiary Age form the nearby La Sal, Abajo, and Henry Mountains.

Second among the main events leading to the formation of the canyon country was the raising and buckling of the Plateau by earth forces so that it could be vigorously attacked by various forces of erosion and so that the rock materials thus pried loose or dissolved could eventually be carted away to the Gulf of California by the ancestral Colorado River. Some idea of the enormous volume of rock thus removed is apparent when you look down some 2,000 feet to the river from any of the high overlooks such as Dead Horse Point or Green River Overlook (**Figures 25 & 26**). Not so apparent, however, is the fact that some 10,000 feet of younger Mesozoic and Tertiary rocks that once overlay this high plateau also has been swept away. In all, the river has carried thousands of cubic miles of sediment to the sea and is still actively at work on this gigantic earthmoving project.

In an earlier report, it was estimated that the rate of removal may have been as great as about 3 cubic miles each century. For a few years the bulk of it was dumped into Lake Mead, but now Lake Powell is getting much of it. When these and other reservoirs ultimately become filled with sediment, for reservoirs and lakes are but temporary things, the Gulf of California will again become the burial ground.

Last but far from least among the factors responsible for the grandeur of the canyon country is the desert climate which allows observations of virtually every foot of the vividly colored naked rocks and has made possible the creation and preservation of such a wide variety of fantastic sculptures. A wetter climate would have produced a far different and smoother landscape in which most of the rocks and land forms would have been hidden by vegetation.

Figure 25. Dead Horse Point State Park provides scenic views of the amount of erosion occurring from uplift of the plateau region and the incision of the Colorado River. Sources: Upper photograph by Utah State Parks posted on the internet. Lower photograph by American Southwest posted on the internet.

Figure 26. Another view from Dead Horse Point State Park displaying the amount of rock removed by erosion throughout the park. Source: Worldwide Elevation Map Finder posted on the internet.

In the canyon lands, the vegetation is mainly on the high mesas and on the narrow flood plains bordering the rivers, but scanty vegetation does grow on the gentle slopes or flats.

The desert climate has combined with the nearly flat lying layers of sediments of different character, hardness, and thickness to produce steep slopes having many cliffs and ledges and generally sharp to angular edges rather than the subdued rounded forms of more humid regions. This has led geologists to refer to such terrain as having "layer-cake geology," and this is brought out by the profile in the rock column, by cross sections, and by many of the other photographs taken in the canyon lands park.

However, because of the lateral changes in thickness and character and the wedging out of certain beds such as the White Rim Sandstone Member of the Cutler Formation, no two sections of the strata were exactly alike. An often-asked question is why are most of the rocks so red? This can be answered by one word-iron, the same pigment used in rouge and in paint for barns and boxcars. Various oxides of iron some including water produce not only brick red but also pink, salmon, brown, buff, yellow, and even green or bluish green. This does not imply that the rocks could be considered as sources of iron ore, for the merest trace of iron generally only 1 to 3 percent is enough to produce even the darkest shades of red. The only rocks in the park that contain virtually no iron are white sandstones of the White Rim Sandstone Member of the Cutler Formation and the Navajo Sandstone.

Microscopic examination of the colored grains of quartz or other minerals shows the pigment to be merely a thin coating on and between white or colorless particles. Sand or silt weathered from such rocks soon loses its color by the scouring action of wind or water so most of the sand dunes and sand bars are white or nearly so. Maps and cross section of the park show that in general the major features of the landscape lie at three different and distinctive levels. A recently erected plaque on Grand View Point appropriately refers to these levels as the "Three Worlds." The high plateaus or mesas in and adjoining the park dominate the skyline. In fact, the central one between the Green and Colorado Rivers is appropriately named Island in the Sky (**Figures 27 & 28**).

Observing to the east or the west shore of this towering cliff-bordered island, one can look across a sea of fantastic erosional forms to a similar cliff-bordered shore at about the same level. Closer inspection of the sea of rocks on either side shows relatively flat benches or platforms about halfway to the bottom. Below these are the generally steep sided or cliff bordered canyons of the two rivers and their larger tributaries. From some vantage points along the shore such as Dead Horse Point or Green River Overlook, the deepest level of all-the channels and flood plains of the Green and Colorado Rivers can be observed (**Figure 29**).

Figure 27. View from the Island In The Sky overlook into Canyonlands National Park and Shafer Canyon from White Rim Road. Source: Roads less traveled posted on the internet.

Figure 28. An alternative view from Islands In the Sky of Canyonlands National Park. Source: Chauna's Adventures posted on the internet.

Figure 29. The Green River Overlook provides views of the deepest level of erosion carved by the Green River. Source: National Park Service posted on the internet.

What caused the "Three Worlds" and the formidable cliffs supporting the high mesas or forming towering monoliths like Angel Arch or Druid Arch? Differences in the composition, hardness, arrangement, and thickness of the rock layers determine their ability to withstand the forces of fracturing and erosion and hence their tendency to form cliffs, ledges, or slopes. Most of the cliff- or ledge-forming rocks are sandstones consisting of sand grains deposited by wind or water and later cemented together by silica (silica oxide, calcium carbonate) or one of the iron oxides but some hard resistant ledges are made of limestone (calcium carbonate). A rock column of the stratigraphy shows in general how these rock formations are sculptured by erosion and how they protect underlying layers from more rapid erosion. The nearly vertical cliffs supporting the highest mesas consist of the well-cemented Wingate Sandstone protected above by the even harder sandstone of the Kayenta Formation (**Figures 30 & 31**).

Vertical cliffs and shafts of the Wingate Sandstone endure only where the top of the formation is capped by beds of the next younger rock unit-the Kayenta Formation. The Kayenta is much more resistant than the Wingate, so even a few feet of the Kayenta protect the rock beneath. In some places remnants of the overlying Navajo Sandstone make up the topmost unit of the cliff.

Figure 30. Angel Arch in the Needles District of Canyonlands are supported by Wingate Sandstone (tan to white) and overlying Kayenta Formation (red). Source: QT Luong posted on the internet.

Figure 31. Druid Arch is composed of Wingate Sandstone. Source: Oros posted on the internet.

Park Observations

The question of how to see the park has no simple answer for the park is too vast and complex to comprehend by a quick visit to any one of its many and varied parts or by any one means of transportation. Some, as did Major Powell, view it only from the rivers-by boat plus a few back-breaking climbs up the bordering canyon walls. Others see only the small parts reachable by passenger cars. The more venture-some see vastly more by jeep, foot, or horseback. And a few prefer to view it as the birds do-from the air. Those who put aside their magazines long enough get bird's-eye views without half trying for Canyonlands is beneath the principal air routes connecting Los Angeles with Grand Junction and Denver. Actually, a full appreciation of all the wonders and beauties of the park is possible only by combining all these approaches and methods of locomotion, but only a few fortunate souls have thus been able to inspect virtually every square foot of it. The task clearly then is how best to present such a complex wonderland to the reader. The method selected after considerable thought and a few false starts is to begin at the top of the high mesas-and work downward much as the rivers have done in carving out this fantastic area to some of the broad bench lands beneath the mesas and eventually to the river channels and deep canyons. Although the approach selected may not be the best, and admittedly is but one of several that comes to mind, it gets the job done.

The High Mesas

Even though the "peninsular" mesas east and west of Island in the Sky known respectively as Hatch Point and the Orange Cliffs lie outside the present boundaries, they provide breath taking views of important features within the park so brief descriptions of them are included below. But first, let's take a closer look at Island in the Sky (**Figures 32 & 33**).

Island in the Sky

Island in the Sky is really a fork of a wedge-shaped peninsula extending southward between the two rivers. An outlier to the south named Junction Butte has already been severed from the main peninsula by erosion and now is a true island. A large chunk of Island in the Sky south of The Neck was about to be severed by erosion from the main peninsula to become a true island when recent widening and grading of the road gave it a temporary reprieve. Furtive glances to right or left showed the two canyons perilously close, and complete severance seemed imminent.

The road builders have staved off disaster for a few thousand years, but ultimately the large section to the south will become another island, and a bridge will be required to connect it to the mainland. Its appearance from the air before the road widening is shown in **Figure 34**.

Figure 32. The Hatch Point overlook into Canyonlands provides views of mesas to the east. Source: Lynn Sessions posted on the internet.

Figure 33. The Orange Cliffs overlook provides views of Canyonland mesas to the west. Source: Touristlink posted on the internet.

Figure 34. Aerial view of The Neck and Shafer Trail leading up into the head of Shafer Canyon. View is facing southwest, after rebuilding of park road on mesa top. Cliff-walled canyon to right of The Neck drains westward to the Green River. The south fork of Shafer Canyon to left drains eastward to Colorado River. This is the narrowest part of Island in the Sky.

The entrance road to Island in the Sky intersects U.S. Highway 163 at a point 10 miles northwest of Moab, or 21 miles southeast of Crescent Junction on Interstate Highway 70. From U.S. 163, a paved road climbs colorful Sevenmile Canyon past sandstone cliffs of the Wingate, Kayenta, and Navajo Formations to reach the high mesa (**Figure 35**). There, just "offshore" to the north are anchored the "battleships" that guard the island- Merrimac and Monitor Buttes. These landmarks are composed of the Entrada Sandstone-the same rock that forms Church Rock at the entrance to the Needles district and that shapes the spectacular arches in Arches National Park. All three members of the Entrada are present here as well as at Church Rock (**Figure 36**). Eleven miles from the junction with U.S. Highway 163, a graded road to the right called Horsethief Trail goes 16 miles down to the Green River where it connects with roads following the river both upstream and downstream. The road upstream leads to two uranium mines in the lower part of Mineral Canyon which were reactivated in 1972 and 1973. The switchbacks are quite spectacular and are reminiscent of the Shafer Trail (**Figure 37**).

Figure 35. Sevenmile Canyon as viewed from Arches Scenic Drive exposes the three formations belonging to the Cutler Group: Wingate Sandstone, Kayenta Formation, and the Navajo Sandstone.

Figure 36. Church Rock at Monticello Utah exposes all three members of the Entrada Formation. Source: Istock posted on the internet.

Figure 37. The Lower Mineral Canyon is a former location for uranium mining in the early 1970's. Source: Youtube posted on the internet.

Three miles south of the Horsethief Trail turnoff is a fork in the road. To the left, the pavement continues to Dead Horse Point, and straight ahead a graded road leads southward to the Island in the Sky district of Canyonlands National Park.

Most of Island in the Sky has a scattered growth of pinon and juniper trees, but several large flat areas such as Grays Pasture contain sufficient sandy soil to support a mantle of grass and weeds which is used for grazing. However, grazing in this part of the park was discontinued in 1975.

Dead Horse Point State Park

Follow the paved road from U.S. Highway 163 all the way to Dead Horse Point which was set aside as a state park in 1957. The park has a visitor center, museum, modern campgrounds and picnic facilities, and piped water which is hauled all the way from Moab. An entrance fee permits a drive across the narrow neck to a parking area near the point proper which is protected by stone walls and is provided with a ramada, benches, paths, and sanitary facilities. From Dead Horse Point there are breath taking views in several directions including a loop of the Colorado River called the Goose Neck, 2,000 feet nearly straight down (**Figure 38**).

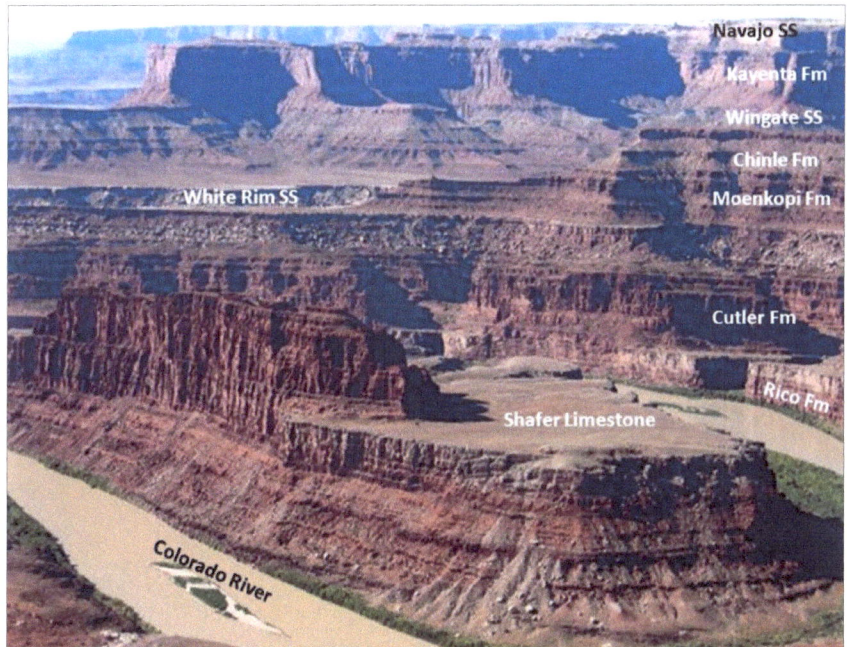

Figure 38. Goose Neck on the Colorado River as viewed from Dead Horse Point State Park. The view provides an excellent view of Canyonlands National Park's stratigraphic column. Source: Utah State Parks posted on the internet.

How did such a magnificent viewpoint get such a macabre name? Dead Horse Point was named for a sad but colorful legend concerning a band of wild horses that once roamed the high mesas. The point is really an embryo island separated from the mainland by a narrow neck barely wide enough for the present road. In the early cowboy days the island was used as a natural corral in which wild mustangs were penned up behind a short fence across the neck so that the better ones could be in Colorado. A band of horses corralled too long without water allegedly died of thirst within sight of the river 2,000 feet below hence the name of the point, or at least so one version of the story goes. Some versions allude to the wranglers as cowboys; others, as horse thieves.

To the northeast, the Cane Creek anticline is an upward fold of the rocks behind which looms the La Sal Mountains (**Figure 39**). A better view of the Cane Creek anticline can be seen from Anticline Overlook. From this vantage point at Dead Horse Point, much of Hatch Point including Anticline Overlook can be observed by looking east and southeast. Spectacular views of the northern part of Canyonlands National Park lie to the south, southwest, and east. Looking southwest, most of the rock formations exposed in Canyonlands may be observed, more than can be seen from any other vantage point in or near the park.

Figure 38. The Cane Creek anticline is viewed from the Anticline Overlook in Monticello. The fold was pushed up by salt deposits flowing upward. The bluish white area in the upper left belong to potash evaporation ponds. Cane Creek runs through the upper center of the imagery.

Parts of Shafer dome is a general domelike shape outlined by the bluish-white Shafer limestone, a marker bed which also caps the bench on the peninsula within the Goose Neck of the river. This limestone which forms the top of the Rico Formation is not shown because its exposure is limited to the Shafer dome and the Cane Creek anticline and its name is used only locally by prospectors for oil and gas.

The White Rim Sandstone Member of the Cutler Formation referred to hereinafter simply as the White Rim Sandstone becomes thinner toward the northeast but is absent entirely just a short distance to the northeast. The gradual disappearance of recognizable beds of this type toward the northeast, including the disappearance of some limestone beds containing marine fossils are examples of what geologists call facies changes. At this location, the changes result from the fact that while strata were being deposited in or near ancient seas that lay to the southwest beds of different character were being laid down on land by streams emanating from the northeast.

North Entrance

The north entrance to the Island in the Sky district of Canyonlands National Park used to be 6 miles south of the junction with the paved road to Dead Horse Point, but since the land additions of November 1971, it is only 4.5 miles south of this junction. A temporary trailer-housed entrance station marks the old boundary.

Shafer White Rim Trail

During the early 1950's, a remarkable but hair-raising road known as Shafer Trail was cut down the face of the cliffs below The Neck to reach the C Group of uranium claims near the head of Lathrop Canyon. It branches southward from the park road a mile south of the new entrance, then descends in a series of switchbacks. The photograph view shows the upper trail (right) and The Neck (left) before the park road was graded and widened, and a view from near The Neck shows the precipitous cliffs the trail descends. It follows the general route of an old foot trail (**Figure 39**).

Figure 39. The Shafer White Rim Trail Road (right side of the photograph) with switchbacks on the descent into Lathrop Canyon. The White Rim Sandstone underlies the road on the right. Source: Only In Your State posted on the internet.

Shafer Trail connects with the White Rim Trail which as the name suggests is built mainly on the White Rim after which the White Rim Sandstone was named. The White Rim Trail can be followed northeastward to join the pavement at Potash, or it can be followed southward along the Colorado River canyons to Junction Butte, thence northward along Stillwater and Labyrinth Canyons of the Green River to and beyond the northern boundary of the park. At Horsethief Bottom, the canyon can be exited by Horsethief Trail to rejoin the paved road leading northward to U.S. 163. At Lathrop Canyon, 8 or 10 miles south of where Shafer Trail meets the White Rim Trail, a branch of the White Rim Trail leads downward to the Colorado River where picnic tables and sanitary facilities are provided. This is used as a lunch stop by some boating groups.

Although some two-wheel-drive cars or trucks have traversed the White Rim and Shafer Trails, they may encounter trouble with deep sand, washouts, or fallen rocks so four-wheel-drive vehicles are recommended. In the summer, these trails should not be attempted without plenty of water, and two vehicles traveling together provide an added margin of safety. All vehicles should carry emergency equipment including a shovel, tow chain or rope, jack, tire tools, and other necessary items. Geologists and uranium prospectors working along the White Rim Trail have obtained good drinking water from small springs that flow from the base of the White Rim Sandstone in many places. After rains, runoff gathers in large potholes in the White Rim Sandstone in some places and affords emergency drinking water. Some potholes occur also in the Cedar Mesa Sandstone in the Needles district.

Grand View Point

About a mile southwest of The Neck, the road crosses Grays Pasture-the widest and flattest part of Island in the Sky (**Figure 40**). The drive over this flat grassland yields not the slightest hint of the awesome cliff-walled chasms on either side of the island. Some 5 miles southwest of The Neck, both the island and the road branch like a Y. At a point 0.4 mile north of the Y, Mesa Trail leads one-quarter mile east to Canyon Viewpoint Arch which frames the Colorado River canyon and the La Sal Mountains (**Figure 41**). This arch, at the very top edge of the cliff is composed of the lower part of the Navajo Sandstone. The only other arch of Navajo Sandstone in or near the park that is known, but of course there may be others.

Taking the branch south of the Y follows the narrow crest of Grand View Point for about 6 miles to the main overlook. About 0.9 miles south of the Y, a short walk to the west over the lower part of the Navajo Sandstone affords a magnificent view of Stillwater Canyon of the Green River including Turks Head (**Figure 42**). Half way to the point is a parking area and overlook from which a spectacular view of canyons cutting the White Rim and of the La Sal Mountains beyond are observed. Note that the White Rim Sandstone which forms the broad bench appropriately named the White Rim is here much thicker than where seen near Dead Horse Point. Three more miles southward takes us to Grand View Point and its nearby picnic area. Though named after the former Grand River some 2,000 feet below, Grand View Point has a double meaning.

From here a truly grand view is presented. Looking down at the standing point is the spectacular Monument Basin cut below the White Rim into the brick-red Organ Rock Tongue of the Cutler Formation. The White Rim Sandstone here is slightly thicker than to the northeast but thinner than to the west because it forms a wedge-shaped body that thickens westward. In the distance southeastward are the Abajo Mountains, just west of Monticello, Utah. The prominent projection on Hatch Point on the left skyline is Needles Overlook from which photographs can be obtained. A close up view of Monument Basin showing Junction Butte and Grand View Point in the background is visible from this point. The slender spire in the foreground has a measured height of 305 feet (**Figure 43**).

Figure 40. Grays Pasture is a broad flat land occurring in the Island In The Sky district of Canyonlands. Small buttes are visible along Grand View Point Road.

Figure 41. Mesa Arch and the view of Canyonlands National Park with the La Sal Mountains in the left background. The arch is composed of lower Navajo Sandstone. Source: Valerie and Valise posted on the internet.

Figure 42. The Green River meanders around Turks Head. The view point is on the lower Navajo Sandstone. Source: Art in Nature posted on the internet.

Green River Overlook

About a quarter mile west of the Y, a left fork of the road goes about a mile and a half to Green River Overlook which provides a superb view of Stillwater Canyon of the Green River, the Orange Cliffs beyond, and the Henry Mountains in the extreme distance. Note that here the White Rim Sandstone is much thicker than in preceding views. The prominent butte enclosed by the loop of the river is known as Turks Head and is better seen from the air. The light-colored band near the base of the cliffs in the background is characteristic of the bleached upper part of the Moenkopi Formation in this part of the park. Petroliferous material or odor generally occurs in this bleached zone and in the basal beds of the Moenkopi (**Figure 44**). The campground just north of Green River Overlook has no water as of 1973, but water from wells in Taylor Canyon will eventually be piped to nearby parts of Island in the Sky.

Figure 43. The Grand View Point displays the Monument Basin and the Organ Rock Tongue of the Cutler Formation occurring below the White rim Sandstone. The mesa in the upper left corner belongs to Hatch Point and the Needles Overlook. The mountains in the right center background belong to the Abajo Mountains west of Monticello, Utah. Source: Flickr posted on the internet.

Upheaval Dome

Five miles northwest of the Y, Upheaval Dome is present. Upheaval dome is one of the most unusual geographic and geologic features of the park. Viewed from the air, it resembles somewhat of a volcanic or meteor crater and has been called such by some. Because beds of salt are known to underlie the park some have suggested that the salt may have thickened and welled upward to form a salt dome, similar to domes along the Gulf Coast. However, only 1,470 feet of salt was encountered in an oil test just east of Upheaval Dome. Although salt may have played a role, Upheaval Dome clearly is not a salt dome with dimensions similar to the Gulf Coast types. It may be related to a mound on the deep-seated Precambrian rocks, but the exact origin of the dome is not clear.

Figure 44. Stillwater Canyon and Green River southwest from the Green River loop of river. Brown material covering nearby parts of the White Rim belongs to the lower part of The Green River Overlook. Orange Cliffs are in the background, and the Henry Mountains lie on right skyline. Turks Head in Moenkopi Formation is in the right center of the photograph.

The central part has the structure of a dome recognized by strata dipping downward away from the middle. A ring like syncline or downward fold in the rock layers surrounds the dome beyond which the strata resume their nearly flat position. The white rock in the bottom of the crater like depression is not salt but jumbled large fragments of the White Rim Sandstone. Surrounding that are slopes of the Moenkopi and Chinle Formations, cliffs of the Wingate Sandstone, a circular bench of the Kayenta Formation, and outer ramparts of the Navajo Sandstone. Upheaval Canyon leads to Stillwater Canyon of the Green River at the upper left (**Figure 45**).

One mile before the road ends, a well-marked foot trail leads to the top of Whale Rock, a prominence on the Navajo Sandstone that forms the outer ring of the dome. At the end of the road another foot trail ascends from the picnic area to the foot of the Wingate Sandstone cliffs around the central part of the dome. The views of the dome from these trails are interesting, but the view is really too close to get a true picture of the unusual feature which is obtainable only from the air. Just west of Upheaval Dome, Bighorn Mesa is connected to Steer Mesa by a neck only 15 feet wide flanked by 300-foot vertical cliffs. When this neck is finally breached by erosion, Bighorn Mesa will be just as isolated and inaccessible as Junction Butte, now cut off from Grand View Point (**Figure 46**).

Figure 45. Upheaval Dome of questionable origins exposes stratigraphic column similar to those located elsewhere in the park. Formation labeling is according to the Utah Geological Survey. Source: Airphoto posted on the internet.

Figure 46. Bighorn Mesa (cliffs) are connected to Steer Mesa (meander bend in river) attests to the erosive capacity of the river to create oxbow lakes which has yet to occur but will eventually happen given enough time. Source: David Oppenheimer posted on the internet.

Hatch Point

The high mesa east of Canyonlands National Park and the Colorado River canyons called Hatch Point contains several vantage points ideally suited for viewing scenic features of the park and adjacent areas. Hatch Point is part of the vast public domain administered by the Bureau of Land Management-a sister agency of the Geological Survey and the National Park Service, all of which belong in the U.S. Department of the Interior (**Figure 47**).

Figure 47. View of Canyonlands National Park from the Hatch Point overlook. Wingate Sandstone occurs below the Kayenta Formation cap rock. Navajo Sandstone lies on top occurring as erosional remnants above Kayenta Formation. Source: Adventr.co posted on the internet.

The Bureau, hereinafter referred to simply as the B.L.M. has made many improvements on Hatch Point including fine roads, two modern campgrounds with sanitary facilities, piped water from wells, and two overlooks with protective fences, benches, paths, sanitary facilities, and ramadas containing panels that describe the features visible from the viewpoints. Because of these improvements, the B.L.M. has appropriately named this area "Canyon Rims Recreation Area."

Geologically, Hatch Point is similar to Island in the Sky. Both are bordered by towering cliffs of the Wingate Sandstone capped by the resistant Kayenta Formation, and rounded remnants of the overlying Navajo Sandstone rise above the otherwise-flat mesa surface in many places.

Access to this high tableland is by a good paved road leading west from U.S. Highway 163 at a point 32 miles south of Moab and 22 miles north of Monticello. About 5 miles west of the highway, Wind Whistle Campground is nestled in an attractive cove of Entrada Sandstone cliffs 16 miles from the highway at an intersection (**Figure 48**). From here it is 7 miles west by paved road to Needles Overlook, and 10 miles north to Anticline Overlook. Like the other high mesas, Hatch Point contains peripheral areas of scattered pinon and juniper trees and large flat grasslands used for grazing. Grain tanks here and there store winter feed for the cattle.

Figure 48. Entrada Sandstone exposed along the cliffs of Wind Whistle Campground in Canyonlands National Park. Source Flickr posted on the internet.

Following the pavement leads to Needles Overlook from which fine morning views of Canyonlands National Park can be seen to the south and west (**Figure 49**). Northwestward, a view 10 miles across the Colorado River canyon to Junction Butte and Grand View Point can be made. The feather edge of the White Rim Sandstone caps the White Rim west of the Colorado River, but the White Rim is absent on the east side of the canyon and in the entire Needles district to the southwest where the important scenic features are carved from the underlying Cedar Mesa Sandstone Member of the Cutler Formation, referred to here-in-after simply as the Cedar Mesa Sandstone. Both these sandstones are missing in the foreground – their place being taken by thin beds of red siltstone, mudstone, and sandstone similar to those that comprise the Organ Rock Tongue. These are additional examples of facies changes mentioned earlier.

Figure 49. Needles Overlook provides an overview of Canyonlands National Park. Source: Adem Benemend posted on the internet.

Canyonlands Overlook

Turning north from the intersection 7 miles east of Needles Overlook, a nearly flat grassy tableland to Hatch Point Campground is crossed. The campground is west of the old road. The new road is west of the campground. About a mile before reaching the campground, a jeep trail heads west then northwest about 5.5 miles to Canyonlands Overlook, a scant mile some 1,400 feet above the eastern border of Canyonlands National Park. This overlook affords fine views of the Colorado River canyons and the eastern shore of Island in the Sky (**Figure 50**).

Two miles north of the campground, a minor drainage leading northeastward into the north fork of Trough Springs Canyon is crossed. The B.L.M. plans a road down this canyon to Kane Springs Canyon 1,100 feet below where it will connect both with a scenic drive to Moab, the lower part of which is paved, and with the jeep trail going west over Hurrah Pass and thence south along the eastern benches of the canyons of the Colorado River to the Needles district of the park. Specimens of blue celestite (strontium sulfate, $SrSO$) and barite (barium sulfate, $BaSO$) in the Cutler Formation were discovered at a point where a sharp bend of this jeep trail crosses a fault, or fracture in the northeast fork of Lockhart Canyon. Farther south, the trail swings west of Lockhart Basin whose center exposes part of a syncline.

Figure 50. About 1 mile south of the Hatch Point Campground, a jeep trail turns off to the west to the Canyonlands Overlook point. The viewpoint is in Navajo Sandstone peering out into reddish Kayenta Formation in the background cliffs.

U-3 Loop

Two and a half miles farther north, or about 1.7 miles south of Anticline Overlook, a short road leads to the west and entirely around a small conical butte of the Navajo Sandstone. This new circular drive has not yet been formally named and is simply called the U-3 loop. It affords splendid views to the west and is equipped with picnic tables. Looking west, a W-shaped loop of the Colorado River, Dead Horse Point on the right skyline, and Island in the Sky on the distant skyline are present. The strata curving over Shafer dome appear in the right middle background (**Figure 51**).

Anticline Overlook

About 1.7 miles to the north takes the viewer to Anticline Overlook for the most sublime views in this part of the area. To the north, a view across the northeast flank of the Cane Creek anticline, an up fold of the rocks can be observed. Hurrah Pass straddles the narrow wall separating the Colorado River and its canyon at the left and from Kane Springs Canyon on the right. The Colorado River appears again in the right background where it leaves Moab Valley. The Kings Bottom syncline, or down fold, seen in the middle distance between the Cane Creek anticline and the Moab anticline exposes a wide area of the Navajo Sandstone.

The ridge on the right skyline composed of the Entrada Sandstone is the Windows Section of Arches National Park, and the left skyline shows faintly the distant Book Cliffs. On the east wall of Kane Springs Canyon just to the right is the Atomic King mine in the Cutler Formation from which uranium ore has been mined at intervals during the last 2 or 3 years (**Figure 52**).

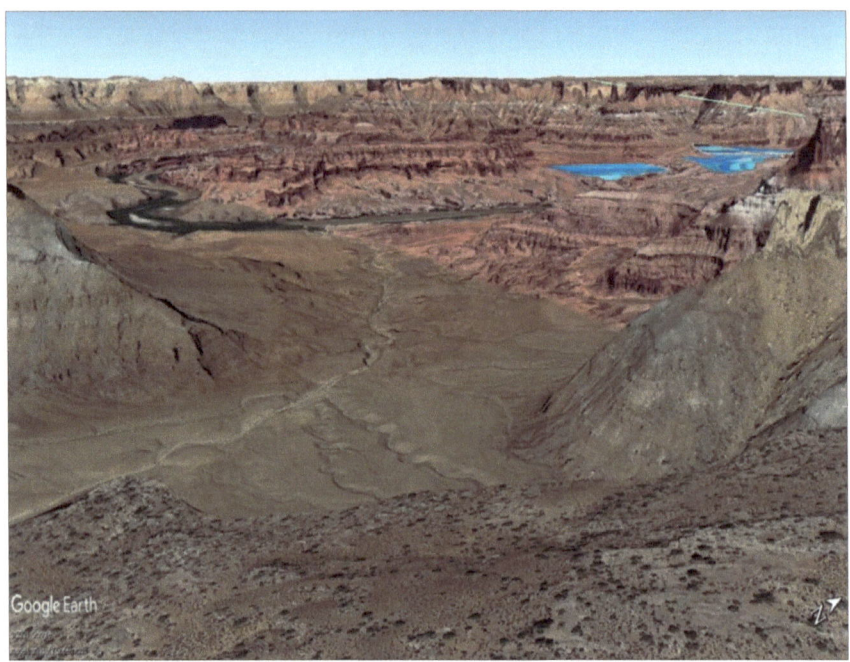

Figure 51. View is westward from above the "U3 Loop" Navajo Sandstone mound described in the text. Dead Horse Point is on the right skyline, Island in the Sky capped by Navajo Sandstone is in the background, Kayenta Formation in foreground on the left. Cliffs topping the ridge on the left are Wingate Sandstone protected by cap rock of the Kayenta Formation. Red slopes beneath Kayenta cliffs are Chinle Formation with the dark ledge of Moss Back Member at the base. Steep slopes and ledges beneath are Moenkopi Formation, the lower part of which is Hoskinnini Tongue. Reddish gentle slopes below are Cutler Formation and the nearly flat benches above Colorado River are Rico Formation with Shafer limestone on top. The blue patches on the right represent potash evaporative basins.

To the northwest is a textbook example of a rock fold-the Cane Creek anticline-laid bare by the Colorado River cutting directly across its crest. Anticlines are noted as sources of or at least hunting grounds for oil and gas, and this one is no exception although production has been relatively small. Some oil and gas was produced also from wells on the north flank of Shafer dome just beneath Dead Horse Point, but other favorable-looking structures farther south that were tested such as Lockhart anticline, Rustler dome, and Gibson dome failed to yield commercial amounts. Exploration for oil and gas led to the discovery of potash beneath several anticlines in eastern Utah and western Colorado.

Figure 52. Anticline Overlook provides a spectacular view of the Colorado River cutting through the central part of the basin. The Cane Creek anticline is to the right of center in the background. On the left, the Hurrah Pass separates the Kane Springs Canyon from the Colorado River. On the right side, the Colorado River exits the Moab Valley at the point where the light colored rocks appear in the photo at the base of the cliff face. Potash evaporation basins appear in the left center on the flanks of the anticline. Source: Where RV Now? Posted on the internet.

The Cane Creek anticline is underlain by about 5,200 feet of salt-bearing rocks in the Paradox Member of the Hermosa Formation of which about 84 percent is halite (common salt, sodium chloride and associated potash salts of sylvite, potassium chloride).

The white area to the left of the mine is waste common salt which is recovered with the potash salts, and the white area with dark stripes at the left is a small part of more than 400 acres of evaporation ponds built to separate the salts. These ponds also can be seen from Dead Horse Point. The dark stripes are the visible parts of plastic membranes lining the ponds. Mining of an 11-foot bed of ore began by usual underground methods from the bottom of a shaft 2,788 feet deep but became too difficult because of intense and intricate folding of the salt beds. Now the salts are being extracted by a method involving solution wherein river water is introduced into the former workings and allowed to stand long enough to dissolve the salts. The brine is pumped out to evaporation ponds, and the valuable potash salts are separated from the sodium salts.

Most of the readily recognizable thin beds such as the White Rim Sandstone pinch out south of here marking the northeastern most exposure of the Shafer limestone at the top of the Rico Formation. Northeast from here, the Rico and overlying Cutler Formation are not readily separable and are included in the so-called Cutler Formation undifferentiated. This land-laid unit of red sandstone, siltstone, and shale is as much as 8,000 feet thick just southwest of the ancient Uncompahgre highland (present Uncompahgre Plateau in western Colorado and eastern Utah) from which it was derived by erosion during the Permian Period.

Orange Cliffs

The high mesas west of Canyonlands National Park do not form as distinct a mainland as does Hatch Point, but rather are broken up into a maze of peninsulas and islands. Owing to the gentle northwestward dip of the rock strata, the altitude of the mesas declines from about 7,000 feet in the south to about 5,300 feet in the north and northwest where the whole aspect of the country becomes more rounded and subdued. The name Orange Cliffs is applied to much of the eastward-facing cliffs which are made of the Wingate Sandstone capped by the Kayenta Formation. Remnants of the Navajo Sandstone increase in number to the north and west where remnants of the next two younger rock units-the Carmel Formation and the Entrada Sandstone-also occur. Thus, the cliff-forming units dip downward beneath younger rocks that form the relatively flat Green River Desert to the northwest, also referred to as the San Rafael Desert (**Figure 53**).

Figure 53. the Orange Cliffs expose remnants of Navajo Sandstone (white) on top of Kayenta Formation (orange cliffs), underlain by Wingate Sandstone (lower most grays).

The areas west of the Green River and the main stem of the Colorado River are the least accessible of any in the park and in this respect have not changed much since Butch Cassidy and his Wild Bunch roamed the area, except that the former main horse trails which are now jeep trails. A secondary road south from the town of Green River goes past the north side of the Horseshoe Canyon Detached Unit and connects with another secondary road to the west which joins Utah Highway 24 at Temple Junction, 20 miles north of Hanksville. Near Horseshoe Canyon, a jeep trail leads south to the Orange Cliffs. Owing to blowing sand, these "roads" are not considered reliable for passenger cars and are best negotiated by four wheel-drive vehicles or horses. The area has changed significantly since the writing of this report. The old trail to Horseshoe Canyon may be difficult to find. Green River is in the Gray Canyon part of Canyonlands north of the national park on Interstate 70.

The Bench-land

The White Rim, a broad bench-land some 1,000-1,200 feet below the southern half of Island in the Sky, and some of the associated bench lands west of the Green River and between the Colorado River and Hatch Point have already been discussed as viewed from Island in the Sky, the White Rim Trail, or Hatch Point. There remains for consideration several other prominent bench lands.

The Maze and Land of Standing Rocks

The Maze, an intricately carved series of canyons and gullies has been called a "Thirty-square-mile puzzle in sandstone", and one can readily visualize a king sized rat struggling in vain to find a way out. The rock is the Cedar Mesa Sandstone which here underlies red shales beneath the White Rim Sandstone. South of The Maze, an area containing tall spires was appropriately named by the Indians *"Toom'-pin wu-near' Tu-weap '*, "or "Land of Standing Rocks" (**Figure 53**).

West of The Maze is Elaterite Basin, so named because of a dark-brown elastic mineral resin called elaterite which seeps from the White Rim Sandstone (**Figure 54**). One of these seeps and a wedge-shaped layer of the sandstone occur in the Range Canyon area was formed by sand being laid down in an offshore bar while red silts and muds were being deposited on land adjacent to the offshore bar. The dark bed just above the White Rim is the Hoskinnini Tongue of the Moenkopi Formation which inter-tongues with and pinches out in beds of the Moenkopi Formation. These are excellent examples of what geologists call facies changes. South of the Land of Standing Rocks are equally colorful areas known as The Fins and Ernies Country. A prominent row of spires near Cataract Canyon is known as The Doll House.

Figure 53. The Maze District of Canyonlands consists of mesa ridges interrupted by narrow canyons. Rock formations consist of Cedar Mesa Sandstone at the base, red shales below White Rim Sandstone on top acting as a cap rock. Source: Bob Rehak Photography posted on the internet.

Figure 54. Elaterite seeping from White Rim Sandstone in Elaterite Basin west of The Maze. Elaterite is a dark-brown elastic mineral resin. Source: Donald L. Baars, posted on the internet by the National Park Service.

The Needles District

The Needles district is currently the most highly developed part of the park as the result of design, not accident, for this district includes the greatest number and widest variety of spectacular features: The Needles proper, The Grabens (pronounced griibans), colossal arches and other erosional forms, large meadows such as Squaw Flat and Chesler and Virginia Parks, a wide variety of prehistoric ruins and pictographs are open to viewing. Confluence Overlook is part of this district where the joining of two mighty rivers-the Green and the Colorado can be viewed. Like the White Rim and The Maze, the Needles district is another of the broad bench lands about midway between the high mesas and the deep canyons (**Figure 55**).

Figure 55. The Needles District from an aerial view. Source: Randall K. Roberts posted on the internet.

Utah Highway 211, as mentioned already, is a 38-mile-long paved road leading to the Needles district from U.S. Highway 163 at a point 15 miles north of Monticello and 18 miles south of La Sal Junction. The intersection is well marked by Church Rock, a butte of the Entrada Sandstone. Highway 211 gradually climbs an eastward-dipping slope of the Navajo Sandstone dotted with a few buttes and patches of the Entrada Sandstone such as Church Rock, and reaches the first of two summits 3 miles west of Highway 163. The road crosses a broad gentle valley in the Navajo Sandstone, reaches the second summit about 10 miles from the highway then descends steeply through the Navajo Sandstone and part of the Kayenta Formation to Indian Creek, 1.5 miles below, and follows this creek nearly to The Needles (**Figure 56**).

Figure 56. Church Rock as viewed from the intersection of US 191 and UT 211. A second butte of similar size is located to the southwest, south of UT 211. Church Rock consists of Entrada Sandstone.

Half a mile down the canyon takes us to the top of the cliff-forming Wingate Sandstone, and another half mile brings us to Indian Creek State Park and its striking Newspaper Rock. Another 1.75 miles takes us to the base of the Wingate and top of the underlying Chinle Formation which forms the red slope beneath the cliffs (**Figure 57**).

Historic Dugout Ranch is 19 miles west of the highway and from here a dry-weather road leads southward up north Cottonwood Creek 37 miles to Beef Basin and connects with roads to Elk Ridge and the Bears Ears, both just west of the Abajo Mountains. Just west of the ranch there is a good view ahead of two historic landmarks: North and South Six-Shooter Peaks so named because of their resemblance to a pair of revolvers pointing skyward. The guns are sculptured from slivers of Wingate Sandstone resting upon conical mounds of the Chinle. These can be seen from a wide area (**Figure 58**).

Southeast of the South Six Shooter Peak on UT 211, the CRC Dugout Ranch is off to the west of the highway. A mile west of Dugout Ranch, the descent passes the top of the Moss Back Member of the Chinle, a ledge of gray-green sandstone forming the base of this generally red formation, then reaches the base of the member at the top of the Moenkopi Formation in the next mile and a half. The Moss Back is uranium bearing in nearby areas. At 3.8 miles west of Dugout Ranch, a poorly marked road on the left crosses Indian Creek, then forks (**Figure 59**).

The left-hand fork follows the bed of Lavender Canyon, and the right-hand fork goes into Davis Canyon. Headwaters of both these canyons are new additions to the park.

Figure 57. From Newspaper Rock State Monument northward, the Wingate Sandstone is exposed along UT 211. The Chinle formation lies on top of the Wingate Sandstone.

The red Organ Rock Tongue of the Cutler Formation is seen about 3 miles beyond the turnoff, or about 6 miles west of Dugout Ranch. Another 1 1/2 miles takes us down in the rock column to the top of the Cedar Mesa Sandstone. The White Rim Sandstone which forms such a prominent bench around the southern part of Island in the Sky and west of the Green River is missing from the Needles district, its place in the rock column being taken by red shales and sandstones of the Cutler Formation. South of Indian Creek other underlying red beds of the Cutler are gradually replaced in turn by the thick Cedar Mesa Sandstone. Erosion has reduced the general level of the Needles district to or into the Cedar Mesa Sandstone, but many streams have cut into the underlying Rico Formation, and the Colorado River has cut also into and in places through the limestones of the unnamed upper member of the Hermosa Formation. The first view of The Needles is another 4 miles, and 1 more mile takes us to the park boundary, nearly 32 miles from the U.S. Highway 163. Highway 163 passes a road on the right leading to Canyonlands Resort, and on the left is a new line camp which replaces the restored one at Cave Spring.

Figure 58. It is difficult to view both south and north peaks together from ground level on UT 211. Traveling on the highway, the south peak is encountered first. To view the north peak, travel further until the highway curves to the left. The peak comes into view on the left.

Figure 59. The greenish gray at the base of the cliff belongs to the Moss Back Member of the Chinle Formation located south of South Six Shooter Peak on UT 211 about one mile west of CRC Dugout Ranch.

The unusual features of the Needles district are due in some part to the character and thickness of the underlying rocks but in large part to erosion along joints and faults. Joints are fractures along which no displacement has taken place, and faults are fractures along which there has been displacement of the two sides relative to one another.

The Cedar Mesa Sandstone comprises 500 to 600 feet or more of hard well cemented buff, white, and pink beds of massive sandstone. On the basis of the type and amount of deformation and erosion of the Cedar Mesa Sandstone and underlying rocks, the Needles district can be divided into three differing areas: (1) an eastern area where the rocks are relatively undeformed but are carved into an intricate series of canyons including Salt Canyon and the upper reaches of Davis and Lavender Canyons- the section of the district that contains most of the arches and Indian ruins; (2) The Needles proper, where tensional forces have cracked the brittle Cedar Mesa Sandstone into a crazy-quilt pattern of square to rectangular blocks separated by joints widened by erosion, creating a myriad of spires and pinnacles; and, (3) The Grabens where the previously jointed rocks were later subjected to additional tensional forces that produced a series of nearly parallel faults that trend northeastward and separate down dropped blocks of rock called grabens (pronounced grabens as in horst and graben) from intervening stationary or up thrown blocks of rock called horsts (**Figures 60, 61, & 62**).

Figure 60. *This image is the eastern Needles District composed of intricate canyons and arches.*

Figure 61. The Needles District proper is the central area where tensional forces have cracked the brittle Cedar Mesa Sandstone into a crazy-quilt pattern of square to rectangular blocks separated by joints widened by erosion, creating a myriad of spires and pinnacles.

Figure 62. The western part of the Needles District belong to The Grabens where the previously jointed rocks were later subjected to additional tensional forces that produced a series of nearly parallel faults that trend northeastward and separate down dropped blocks of rock called grabens (pronounced grabens as in horst and graben) from intervening stationary or up thrown blocks of rock called horsts.

For traveling to most features, a four-wheel-drive vehicle is strongly recommended. Some visitors negotiate the jeep trails with dune buggies or motorcycles, but four-wheel-drive vehicles are considered safer and generally more reliable. A few trails can be traveled only on foot.

Squaw Flat, in the western part of the relatively undeformed area is a nearly flat area of lower Cedar Mesa Sandstone covered here and there by a thin layer of sparsely vegetated soil and surrounded by generally low hilly erosional forms in the upper part of the sandstone. Short canyons and alcoves in the sandstone hills along the west side afford excellent semi-private campsites each of which has its own paved access road, picnic table, and trash can. Moreover, ground water at shallow depth in the underlying sandstone has encouraged the growth of exceptionally large pinon and juniper trees that provide welcome shade (**Figure 63**).

Figure 63. The Squaw Creek Campground is surrounded by Cedar Mesa Sandstone. View point is from Loop A above the campground. Source: Posted on the internet.

Salt, Davis, and Lavender Canyons

A glance at the southeast corner of the Canyonlands route map shows that most of the arches and prehistoric ruins in the park are in Salt Canyon and its main tributary, Horse Canyon. A few are in adjacent Davis and Lavender Canyons whose headwaters were recently annexed to the park. These canyons are accessible only by negotiating the streambeds on four-wheel-drive vehicles, horseback, or foot.

Salt or Horse Canyons are best conquered by four-wheel-drive vehicles plus short hikes in the northern part and long hikes in the southern part. An aerial view eastward across Salt Canyon shows that erosion has produced an intricate series of meandering canyons separated by rather narrow walls of the Cedar Mesa Sandstone resembling somewhat The Maze west of the Green River (**Figure 64**).

Figure 64. Salt Creek Canyon branches off the Colorado River below the confluence with the Green River at The Loops. The narrow canyons are composed of Cedar Mesa Sandstone along Salt Creek.

The massive sandstone beds of the Cedar Mesa are composed of sand grains cemented together by calcium carbonate ($CaCO_3$) which also forms the mineral called calcite and the rock known as limestone. Limestone and calcite are soluble in acid, even weak acid such as carbonic acid ($H \cdot HCO$) formed by solution of carbon dioxide (CO_2 and water). Ground water, found everywhere in rock openings at differing depths beneath the surface contains considerable dissolved carbon dioxide derived from decaying organic matter in soil from the atmosphere and from other sources. Even rain water and snow contain small amounts absorbed from the atmosphere, enough to dissolve small amounts of limestone or of calcite cement in sandstone. The calcite cement in the Cedar Mesa and many other sandstones are unevenly distributed so the cement is removed first from places that contain the least amounts, and once the cement is dissolved, the loose sand grains are carried away by gravity, wind, or water. Thus, relatively thin walls of sandstone containing irregularly distributed patches of soluble cement are prime targets for the formation of potholes, alcoves, and caves.

Once a breakthrough occurs, weakened chunks from the ceiling tend to fall off, and arches of various shapes are produced because an arch is naturally the strongest form that can support the overlying rock load. All the spectacular arches about to be seen were carved from the Cedar Mesa Sandstone.

Let's begin the tour of Salt and Horse Canyons by driving a four-wheel-drive vehicle eastward from the fine campground at Squaw Flat (the road is paved as of this writing). After about a mile the road passes the Wooden Shoe, capping a ridge south of the highway. It contains one of the smallest arches encountered (**Figure 65**).

Figure 65. Wooden Shoe Arch is one mile east of the Squaw Creek Campground forming in Cedar Mesa Sandstone. Source: Utah Arches & Bridges posted on the internet.

Three quarters of a mile east of the temporary ranger station we come to Cave Spring, an old restored cowboy line camp. This and an adjacent cave containing a spring is part of the interesting well marked Environmental Trail well worth the half hour or so it requires (**Figure 66**).

The jeep trail up Salt Canyon lies mostly in the sandy bed of Salt Creek but includes a few shortcuts across goosenecks and some rough rocky stretches around rapids or waterfalls. It is best traveled when the canyon bottom is moist but not soaked. When the sand is soft and dry, a shift into four-wheel drive is generally necessary. Signs warn of quicksand which occurs when the sand is fully saturated. Hence, summer thundershowers sometimes require delaying or postponing the trip.

Figure 66. Cave Spring exposes layered bedding in the Cedar Mesa Sandstone. Source: TripAdvisor posted on the internet.

When in doubt, consult a park ranger for expected weather and trail conditions. Thundershowers sometimes occur so suddenly and violently as to cause serious floods, and the "road" is closed when heavy rain is expected. However, if an unexpected storm occurs while you are up in the canyon, try to reach high ground and wait until the flood subsides. If you do not have time to get your vehicle out of the flood's path at least get yourself and passengers to a safe spot.

Two and a half miles south of Cave Spring, the confluence with Horse Canyon is reached marked by a sign at the Y giving distances to points of interest up each canyon. Let's try Horse Canyon first. After about a mile, Paul Bunyan's Potty is on the left-one of the most aptly titled features of the park (**Figure 67**). Two miles south of the Y is Keyhole Ruin, nestled in a cleft high on the cliff to the left-a granary built by the Anasazi. At this point is another Y. The left fork leads half a mile east to Tower Ruin, one of the largest and best preserved Anasazi granaries in the park. The right fork takes us on up Horse Canyon, and in about 2 miles the road passes Gothic Arch on the right (**Figure 68**). In 2 more miles, 4 miles from Salt Canyon, a short hike up the tributary to the right leads to Castle Arch and Thirteen Faces. Assuming photographs of the important features were taken along the way, it probably is about time to return to camp at Squaw Flat unless the choice is to spend the night at Peek-a-boo Spring and primitive campground in Salt Canyon, about 1.2 miles above the confluence with Horse Canyon.

Figure 67. Paul Bunyan's Potty is a pothole occurring at the confluence of Salt Canyon and Horse Canyon. Calcite carbonate cement in the sandstone dissolved out leaving a pothole in the rock for runoff to discharge into the narrow canyon below.

Figure 68. Gothic Arch (right) and Little Gothic Arch (left) appear at the top of the mesa about 2 miles on the south side of the road when entering Horse Canyon.

Another drive takes a trip up Salt Canyon 8.5 miles past the confluence with Horse Canyon to another confluence and Y which has a primitive campsite without water. One mile up the left, or southeast tributary is a parking area where a hiking trail begins the 0.5 mile walk to Angel Arch considered by many people to be the most beautiful and spectacular arch in the park if not in the entire canyon country (**Figure 69**).

Figure 69. Angel Arch can be viewed from a trail leading up a tributary canyon from Salt Canyon. Source: TripAdvisor posted on the internet.

From the last Y, proceed only about 2.5 miles farther up main Salt Canyon by vehicle, and the remaining features can be reached only on foot. The All American Man, a unique pictograph referred to earlier is about 3.5 miles up the canyon. Those hardy souls who wish to hike many additional miles to the head of Salt Canyon will be rewarded with views of four additional arches and several ruins. The more adventuresome may wish to explore upper Lavender and Davis Canyons by driving up the sand washes in a four wheel drive vehicle, but inquiry should be made from a park ranger regarding access to the canyon mouths and condition of the washes. Hand Holt Arch and Cleft Arch are two of the rewarding sights in Lavender Canyon.

The Needles and the Grabens

The northeastern edge of The Needles proper can be seen from Squaw Flat, but the true character of The Needles can be appreciated better from the air. One cannot get far into The Needles without traversing part of The Grabens, so both features will be considered together. Hiking into The Needles and The Grabens from Squaw Flat is possible, but let's make the trip using a four-wheel-drive vehicle and several short hikes.

Ordinary passenger cars now can go 2.75 miles west of Squaw Flat to Soda Spring, at the east base of Elephant Hill, but beyond Soda Spring four-wheel-drive vehicles should be used. Some people conquer the hill in dune buggies or on motorcycles, but this is considered quite dangerous. Both sides of this short (1.5 miles) but formidable hill has switchback curves too sharp to negotiate in the regular manner, so special driving techniques must be followed. On the east side, you must drive out on a flat rock, jockey back and forth until turned completely around, then proceed up the hill. On the west side, the descent is a 40-percent grade to a shelf, *back* down a narrow stretch of about 30-percent grade and back sharply to the left onto a flat rock, then go forward again. On the return trip the whole procedure is carried out in reverse order (**Figure 70**).

Figure 70. Elephant Hill access is limited to the locations described in the text. This road is blocked and is only available for hiking access only. Source: Geogypsy posted on the internet.

West of Elephant Hill, the road reaches a Y. Turn left on a one-way road. The right-hand road is for later one-way return to the Y. Why the left-hand fork is one way soon becomes apparent, for the road leads into a narrow shallow graben called Devils Pocket between rock walls, and is barely wide enough for one car. After about 2 miles the graben widens out into a beautiful spot called the Devils Kitchen which contains several picnic tables tucked into shady recesses in the sandstone walls. This is the starting point for two trails leading southward by different routes to Chesler Park from which other trails lead to Druid Arch or back to Squaw Flat (**Figure 71**).

Figure 71. Devil's Pocket is a narrow elongated canyon that leads into the Devil's Kitchen (top of image) where a primitive campground is present. Devil's Pocket is a graben down faulted between horsts.

From the Devils Kitchen, the road turns abruptly westward for about half a mile to another Y in about the middle of Devils Lane, one of the larger grabens and one of two whose entire length is traversed by roads (**Figure 72**). Only the left fork is a two-way road, so let us take the left fork 2.75 miles southwestward to the next road junction. About halfway down Devils Lane, a fault crossing the graben has created a narrow steep ridge appropriately called SOB Hill, because the road over it creates a challenge that some vehicles fail to meet on the first attempt!

Figure 72. Devil's Lane opens up from Devil's Kitchen campground heading off to the north.

The next road intersection is a sharp turn leading southwest to Ruin Park and Beef Basin. The abandoned left fork leads east into Chesler Park. This park is a beautiful natural meadow of several hundred acres fenced by a natural wall of needles and containing a central island of needles. Because of vehicular damage to meadow vegetation, the National Park Service found it necessary to close the road. To reach Chesler Park now, vehicles must go right a short distance to the Chesler Canyon turnoff, then left about half a mile to a parking area. From here, a 0.5 mile hike east through the narrow Joint Trail gets us to the south side of Chesler Park where the trail joins the abandoned road to reach the northeast corner of the park and the trails into The Needles proper (**Figure 73**).

This change adds 1.75 miles (one way) to the hike to Druid Arch making the round trip about 11.5 miles. At the old trailhead near the northeast corner of Chesler Park is a sign proclaiming the need for rubber-soled shoes and water, and it is strongly supported that these items are required for much of the hike is on bare smooth sandstone and includes steep slopes and *generally* dry waterfalls. The hike should not be attempted by anyone not in good physical condition, and it should not be undertaken alone. Two or more people should travel together.

The trail to Druid Arch from Chesler Park starts out on bare Cedar Mesa Sandstone marked by a succession of rock cairns two of which are visible and without which the trail would soon be lost. The trail drops rapidly down into Elephant Canyon which is then followed southward 2 miles to the arch (**Figure 74**).

Figure 73. Chesler Park is the entry way into the Needles proper (left background). Cedar Mesa Sandstone makes up the lower part of the park and is composed of narrow canyons and dry waterfalls. The Needles has jointing patterns which are barely wide enough for a single person to pass through on the Joint Trail. Source: Grand Canyon Trust posted on the internet.

Figure 74. Druid Arch with rock cairns (stacks of rock) in the foreground and Needles in the background. Cedar Mesa Sandstone in the foreground. Source: Posted on the internet.

Elephant Canyon cut through the Cedar Mesa into the underlying Rico Formation, and much of the canyon is quite narrow and steep sided. Although much of the Rico consists of red beds laid down above sea level by ancient streams, the trail crosses several thin beds of dark-gray hard limestone containing fossil marine seashells and ancient sea anemones whose original calcium carbonate parts have been locally replaced by jasper (red iron-bearing silica). When at last the weary hiker makes the steep climb out of the canyon and rounds the final bare-rock curve, the sudden and striking view of Druid Arch seems worth every bit of the effort.

After hiking to Druid Arch and after the length of this route was increased to a round-trip distance of 11.5 miles, a new route was constructed having a round-trip length of only 8.5 miles. This new trail starts at the end of the passenger-car road at the east edge of Elephant Hill, goes 1.25 miles southwest to join an older trail in Elephant Canyon, then follows this canyon 3 miles south to the arch.

After returning to the vehicle west of Chesler Park and backtracking over SOB Hill to the intersection in the middle of Devils Lane, proceed northward on a one-way road to and beyond the Silver Stairs for a closer look at Devils Lane and other grabens to the west and for a look at the confluence of the Green and Colorado Rivers. But first, pause and reflect upon the possible origin of The Grabens (**Figure 75**).

Figure 75. SOB Hill is the narrow canyon at the red marker. To the north of SOB Hill, the Rico Formation is exposed in the flatlands as the pinkish red unit underlying the Cedar Mesa Sandstone (white unit) on the floor of the flat. The gray area to the left is Devil's Lane. At the top of Devils Lane is Devil's Kitchen campground (left of center).

Geologists have different opinions as to just how grabens and complex systems of joints have formed, but all seem to agree that tensional forces were involved. Some think that solution of salt and gypsum from the Paradox Member of the Hermosa Formation by ground-water movement allowed the brittle Cedar Mesa Sandstone and other overlying rocks to sag, producing tension cracks and faults. Others believe that removal of the salt and gypsum occurred by plastic flowage toward the Meander anticline whose axis follows the Colorado River southwest from The Loop past the confluence, and to and beyond Spanish Bottom. Some suppose that compaction due to the weight of the abnormally thick pile of sedimentary rock underlying the area may have caused the sagging, cracking, and faulting. The rock deformation may have resulted from a combination of these and possibly other things, of course but whatever the cause the resulting features are very striking. The tension faults are called normal faults in contrast to faults formed by horizontal compression which are called reverse faults (**Figure 76**).

Figure 76. The Meander Anticline is marked by the black dashed lines. Beds to the left are tilting southeast, and the beds on the right are tilting northwest.

The Grabens range in width from about 7 or 8 feet at the north end of Devils Pocket to nearly 2,000 feet at the south end of Red Lake Canyon, but the average width is about 500 feet (**Figure 77**). The floors of The Grabens are covered by soil and grass, but the displacement along the faults is believed to approximate the height of the walls-nearly 300 feet.

Figure 78. Red Lake Canyon (aka Devil's Lane) is a faulted graben dropped down between two horsts. The cliff walls represent the faults. Source: American Southwest posted on the internet.

That The Grabens are of fairly recent origin is attested by the fact that most of the walls are vertical fault faces showing little sign of erosion, and that no through drainage has yet been established in Cyclone Canyon which is a series of basins with low divides between. Several pre-existing streams were interrupted or diverted by the faulting (**Figure 78**).

Continuing the journey northward along Devils Lane just before reaching the Silver Stairs, the visitor may wish to pause long enough to take in the distant view to the northwest toward Junction Butte and Grand View Point. After descending the steep Silver Stairs in a narrow cleft between rock walls, another intersection is reached involving a two-way road continuing northwest which arrives at the next destination, and a one-way road turning right returns to Elephant Hill via part of Elephant Canyon.

About 2 miles to the northwest the north end of Cyclone Canyon crosses the largest graben. It contains a road 3.5 miles long and is well worth seeing. About one-half mile from the south end, an old trail follows Red Lake and Lower Red Lake Canyons to the Colorado River across from Spanish Bottom.

From near the north end of Cyclone Canyon, drive west three-fourths mile to a parking area and hike one-half mile to an overlook for a spectacular view of the confluence of the Green and Colorado Rivers and of the northern part of Cataract Canyon. These and other canyons are discussed in the next chapter.

Figure 79. Cyclone Canyon is adjacent to Red Canyon and is similar in that the cliff walls represent faults and the basin is the down dropped portion of the graben while the cliffs represent horsts. Source: Posted on the internet.

Canyons of the Green and Colorado Rivers

Two of the three major topographic divisions of the park: the high mesas and the bench lands were presented, and there remains to consider the third division: the canyons of the mighty Green and Colorado Rivers and some of their tributaries. A discussion of a few features common to both rivers then the details of each river follows. Above the confluence, both rivers are very crooked and contain many loops, or meanders, the most striking of which are Bowknot Bend of the Green River several miles north of the park, and The Loop of the Colorado River. In contrast, the main stem of the Colorado River below the confluence is considerably straighter. The crooked rivers above the confluence have very gentle grades and are free from rapids or falls whereas a few miles south of the confluence the main stem plunges into Cataract Canyon, the steepest and wildest reach of the river containing 64 rapids (**Figure 80**). These differences are partly explained on the basis of the geologic structure and character of the rocks through which the rivers have cut. Above the confluence, the soft strata dip gently northward. Flowing generally southward, the two rivers are cutting "against the grain" which tends to impede their flow and thus reduce their grades. Below the confluence, the hard limestones of the Hermosa Formation lie relatively flat for several miles and then begin to dip gently southward, thus allowing the river to cut "with the grain" and therefore drop more rapidly.

Figure 80. Section of Cataract Canyon displaying rapids. Hermosa Formation limestone is tilted as exposed at the left footwall which provides a rough surface for the Colorado River to flow over. Source: Cara Befort posted on the internet.

The quiet smooth waters above the confluence permit power boating between the towns of Green River, Utah, and Moab during part of the year whereas the rapids below Spanish Bottom, 3.5 miles below the confluence restrict river travel to float trips using sturdy boats or rafts.

Above the confluence, a so-called Friendship Cruise is run each year during the Memorial Day weekend. Participants tow their own power boats on trailers to the town of Green River, and after the boats are launched, facilities are available at nominal cost for transporting cars and boat trailers to Moab to await the arrival of the boats. Although some high-powered speedboats are reported to have made the run down the Green River to the confluence then up the Colorado River to Moab in a few hours, the trip for most boats requires 2 to 7 days.

Trips by power boats including jet boats can be arranged from either Green River or Moab. Some passengers from Moab return by jeep from Lathrop Canyon via the White Rim Trail, and some from Green River return on land via the Horse-thief Trail. Many prefer the quieter float trip down to the confluence with return to either town by a prescheduled power boat, and some more adventurous souls float through the rapids of Cataract Canyon all the way to Lake Powell.

Entrenched and Cutoff Meanders

Meanders, such as those above the confluence generally are formed by streams flowing in soft alluvium consisting of clay, silt, and sand such as along the Mississippi River below Cairo, Illinois. But there is no soft alluvium along the Colorado and Green Rivers, so how did these meanders form? They probably attained their serpentine shape while cutting in softer younger material which long ago was removed by erosion, and then continued to cut their crooked channels down until they created the deep rock-walled canyons in which they now flow as "entrenched" meanders (**Figure 81**).

Figure 81. Green River meanders are thought to have formed in softer materials that were stripped away by erosion leaving behind entrenched meanders. The red marker is Bowknot Bend. Wingate Sandstone is exposed in Bowknot Bend.

Meandering streams tend to shorten their lengths from time to time by cutting through narrow walls between adjacent loops leaving abandoned horseshoe-shaped channels or lakes. In most of the United States these are known as oxbows or cutoff meanders, but in the desert Southwest they are commonly called by the Spanish term "rincon." Cutoffs are common along soft alluvial channels such as the lower Mississippi River valley but are rare along channels whose meanders are entrenched into hard rock. Thus, there have been many natural (and several manmade) cutoffs along the lower Mississippi during historic times, but the most recent ones along the Green and upper Colorado Rivers probably occurred a million or so years ago, during the Pleistocene Epoch when rivers had high discharge and velocity.

Labyrinth Canyon was named for its deeply entrenched meanders. Nevertheless, the distant future will likely see a breakthrough whereby Green River will shorten itself by about 7 miles. It is interesting to note that the vertical cliffs of Wingate Sandstone in and west of Bowknot Bend are only a few hundred feet above the river whereas because of the gentle northward dip of the beds and the gentle southward grade of the rivers, the Wingate cliffs are more than 2,000 feet above the two rivers at Grand View Point and Junction Butte, at the southern tip of Island in the Sky (**Figure 82**).

Figure 82. Junction Butte is composed of Wingate Sandstone. The mesa to the right belongs to Island in the Sky.

At the mouth of Horseshoe Canyon about 3 miles below Bowknot Bend, a large rincon where the Green River shortened its course by about 3 miles occurs. Some idea of the rincon's antiquity is gained from the facts that the river is now some 350 feet lower than at cutoff time whereas Bowknot Bend has shown no visible deepening in 97 years. This rincon was not noted because the banks at this point were not evaluated in the past, but it is quite noticeable on modern topographic maps and on aerial photographs changes occurred. This rincon and Jackson Hole along the Colorado River may be as old as late Tertiary.

At a point reported to be 350 yards above the mouth of Hell Roaring Canyon which enters from the east about 3.5 miles below the rincon indicated that the inscription present at this location is carved on a massive Moenkopi sandstone bed about 40 feet above the canyon floor. A similar inscription was found in Cataract Canyon, 31 miles below the confluence, but it is now covered by Lake Powell (**Figure 83**).

Figure 83. Hell Roaring Canyon exposes Moenkopi Formation along the canyon base. The Chinle formation and Glen Canyon Group overlie the Moenkopi. Source: Mountain Project posted on the internet.

Some boaters are met by car and taken out to Moab or Green River via the Horse thief Trail just north of the park. The road along the river here continues south for 6.5 miles to the mouths of Taylor and Upheaval Canyons where it becomes the White Rim Trail (**Figure 84**).

Coming down the Green River, the river enters Canyonlands National Park where the Grand-San Juan county line meets the Emery-Wayne county line, about 2.25 miles north of Taylor and Upheaval Canyons. The National Park Service had three successful test wells put down in Taylor Canyon, and water under artesian pressure was found in the White Rim Sandstone at depths of 373 to 482 feet.

About 5.5 miles below Upheaval Canyon is an interesting ruin on a hill in the middle of a large nearly closed loop of the river enclosing Fort Bottom. At about the mouth of Millard Canyon leaving Labyrinth Canyon, the Stillwater Canyon is entered. The beginning of Stillwater Canyon is marked by vertical walls of the White Rim Sandstone. A butte to the southwest thought to resemble a fallen cross was named "Butte of the Cross." Farther downstream, instead of a single butte, two buttes appear: a small one in front of a larger one. The feature was renamed "Buttes of the Cross" (**Figure 85**).

Figure 84. Taylor Canyon is a broad canyon exposing Moenkopi, Chinle, and Wingate Sandstone. Chinle Formation rocks are the cliff formers and the Wingate Sandstone caps the cliffs. Source: AllTrails posted on the internet.

About 2 miles below the mouth of Millard Canyon, at Anderson Bottom, one of the most interesting features on the river is reached. The most recent rincon of a major river in the park, if not in the entire canyon country is present. Although some rincons are more recent, they are along minor tributaries such as Indian Creek. The cutoff at Anderson Bottom probably took place during the Pleistocene Epoch whereas most of the others along the main rivers probably occurred during the Tertiary Period. This feature was noted and correctly interpreted as a sharp new course the river took after the cutoff called Bonita Bend.

Continuing through Stillwater Canyon, Turks Head is passed when heading for the confluence of the Green River with the Colorado River (**Figure 86**). It is the canyon just west of the confluence. The lowest and largest cliff above the river is the upper member of the Hermosa Formation overlain by the slopes and thin ledges of the Rico Formation. The massive sandstone at the top of the canyon wall is the Cedar Mesa. Junction Butte and Grand View Point are on the right skyline. The confluence and Cataract Canyon was viewed from the land and from the air. Both features will be described from the Colorado River in the next section.

Figure 85. The Green River's Stillwater Canyon exposed White Rim Sandstone as cliffs. The buttes in the left background belong to the "Buttes of the Cross", southwest of the canyon. Source: YouTube posted on the internet.

The Colorado River

Although the Colorado River enters Canyonlands National Park about 33 river miles below Moab, most boaters or floaters begin their voyage either at Moab or near Potash, and most travelers of the White Rim Trail begin at Moab. This trip starts at Moab. Above the confluence, both the Green and Colorado Rivers are very crooked.

The Kings Bottom syncline southwest of Moab Valley brings the Navajo Sandstone down to and slightly below water level whereas at The Portal, the Navajo caps the southwest wall of Moab Valley (**Figure 87**). Several anticlines at or near the river from Potash to and beyond the confluence bring up strata as old as the Rico or the unnamed upper member of the Hermosa. Between these extremes, much of the river's course lies in strata of the Cutler Formation.

Figure 86. Turks Head exposes the lowermost Hermosa Group overlain by ridge forming Rico Formation. The butte is at the confluence between the Green River and Colorado River. Source: Bob Newitt Photography posted on the internet.

Figure 87. The southwest wall of the Moab Valley displays several interesting components. The Navajo Sandstone caps the cliff. To the left, the Wingate Sandstone underlies the Chinle Formation. The upper Chinle Formation is faulted at the surface. A fault separates the left formations from the right where the right side exposes Moenkopi formation red beds against the Chinle Formation.

About 7 miles below The Portal, Highway 279 is joined on the right by a branch line of the Denver and Rio Grand Western Railroad completed in 1962 to haul potash 36 miles from the mine at Potash north to the main line at Crescent Junction. The railroad emerges from a tunnel at the head of Bootlegger Canyon. Two natural arches near the mouth of the tunnel - Pinto and Little Rainbow Bridge - can be reached by trail (**Figure 88**). About 3 miles farther down the Colorado is a temporary dock from which jet boats and the *Canyon King,* a 93-foot 150-passenger stern-wheeler take off for points downriver during the spring and early summer when water depth permits. The *Canyon King*, a small replica of a Mississippi River stern-wheeler carries passengers about 30 miles downriver to the foot of Dead Horse Point and returns.

Figure 88. The two arches described in the text as Pinto and Little Rainbow Bridge are present and visible along the Corona Trail. The two arches are called Corona and Bowtie by the photographer. The arches shown in this photograph may be two different arches than those described earlier in the caption. Source: Colorado Guy posted on the internet.

About 12 miles below The Portal, the Potash "mine" of Texas Gulf is present. Travelers down the jeep trail below Potash pass the evaporation ponds used to separate the potash from common salt. This segment of the Colorado River has very low grades, and is free from rapids. As with the Green River, the soft rocks along the Colorado have a generally low northward dip that partly explains the river's gentle grade and its southward flow through increasingly lower and older strata. Unlike the Green, however, the gentle dips of the strata in the canyons of the Colorado are interrupted by several gentle anticlinal and synclinal folds and by at least one fault.

The most important of these geologic structures and other features will be noted as the journey down the river progresses.

The first 14 miles from Moab Valley to Potash can be made either by river or by paved Utah Highway 279. This highway leaves U.S. Highway 163 near the uranium ore-reduction plant several miles northwest of Moab, travels through Moab Valley by way of The Portal, and follows the west bank of the river. A paved secondary road from Moab follows the east bank of the river through The Portal and through Kings Bottom where it crosses the Kings Bottom syncline to the mouth of Kane Springs Canyon, then becomes a gravel road that ascends this canyon southward to and beyond Hurrah Pass. High above this road north of Kings Bottom are petroglyphs and a few cliff dwellings in the vertical cliffs of Wingate Sandstone. A ranch "house" at Kings Bottom has been excavated entirely into the Wingate cliff. Convenient turnouts have been provided at several places along Highway 279 for viewing petroglyphs or other points of interest. Small viewing tubes welded to vertical steel posts have signs helping visitors locate and view the features (**Figure 89**).

Figure 89. View from above Moab of the Kings Bottom syncline. Wingate Sandstone cliffs (tan) in the upper part of the image are faulted against red Moenkopi formation in the foreground. Chinle Formation caps the Wingate cliffs.

Across the river east from Potash is Jackson Hole, a large rincon. Since abandonment which shortened the river by about 3.5 miles, the river has cut its channel nearly 200 feet deeper. It is comparable in size to the large rincon along Green River below Bowknot Bend but probably is somewhat younger.

Both rincons may be as old as late Tertiary. Just below Potash, the axis of the huge Cane Creek anticline is crossed (**Figure 90**). The road also leaves Grand County to enter San Juan County. A mile east of this point, high on the canyon wall is the School Section 13 uranium mine which has yielded considerable ore resuming production sometime during 1973. It can be seen from the river or the trail, and some of the tailings are visible on the left flank of the anticline. Voyagers who cross the axis of the Cane Creek anticline may observe on the right-hand (west) bank a protruding oil-well casing, some drill bits, and several shacks-all that remain of the Frank Shafer No. 1 oil test started during the winter of 1924-25 and completed by the Midwest Exploration Co.

The evaporation ponds are in Shafer Basin, a synclinal basin separating the Cane Creek anticline and Shafer dome, as the route crosses the axis of Shafer Basin about 2 miles below the county line. Further downstream is Shafer dome, a closed anticlinal bulge just beyond the W-shaped bend in the river. From almost anywhere in the Goose Neck, the sharp bend of the river, an excellent view of Dead Horse Point some 2,000 feet above is observed.

Figure 90. The Cane Creek anticline is exposed near the Potash (white dashed line). The cliff face behind the dashed line shows the bowing of the fold. The colored beds on the left edge of the image belong to potash beds operated by Texas Gulf.

A small petrified forest resembling a log jam in the eastern part of the Shafer dome at mileage 39 just north of this point about half way between the river and the jeep trail below Dead Horse Point is present. There probably are 20 to 30 logs some of which are as large as 18 inches in diameter and more than 20 feet long, along with a stump about 3 feet in diameter. They occur in red beds at about the middle of the Rico Formation, hence could be either Pennsylvanian or Permian in age. The original wood has been replaced by silicate (SiO_2) and stained a dark reddish brown.

Many teeth of a primitive shark like fish in the Rico Formation were discovered at the same general locality as the petrified wood and also in the Rico on the Cane Creek anticline. The teeth were reported to be one tooth of the cochliodont shark *Deltodus*, and one tooth of the petalodont shark *Petalodus*. About 4 miles below the Goose Neck, we enter Canyonlands National Park and remain in the park almost to the north end of Lake Powell. About 6.5 miles into the park, at the north end of a bend much like the Goose Neck is the mouth of Lathrop Canyon where many boaters stop for lunch and where a side road connects with the White Rim Trail.

Six and one half miles below Lathrop Canyon is the mouth of Rustler Canyon which is joined near its mouth by Indian Creek. The creek follows the highway leading to The Needles from U.S. 163. Within an airline distance of only 3 miles, the lower reach of Indian Creek, an intermittent stream, flows past four small rincons, three of which are within an airline distance of only 0.8 mile. The stream has cut its new channel into the red sandstones and shales of the Cutler Formation only 15 to 20 feet deeper than the abandoned ones in the two rincons and only about 25 feet deeper than the one on the right. This suggests that these cutoffs probably occurred sometime during the Holocene Epoch, or age of man that is probably within the last 10,000 years. A detailed study of these rincons might change this estimate, particularly if, say, buried driftwood or other carbonaceous material could be found for an age determination by the radiocarbon method.

About 5 miles below the mouth of Rustler Canyon and Indian Creek, and also about 5 miles above the confluence is The Loop- an even sharper and more symmetrical figure eight than Bowknot Bend of the Green River. An aerial view of The Loop shows that the channels on the south loop are only about 500 feet apart and that those on the north loop are only about 1,700 feet apart. At the narrowest places both saddles are considerably eroded. The southern one is only about 150 feet above the river, but the northern one is still about 350 feet above. Erosion of both saddles has been hastened by the fact that the axis of the Meander anticline passes through each saddle and that an interesting reverse fault passes through the lower and thinner southern saddle. It seems inevitable that someday the small saddle will be cut through by the Colorado River, and a new rincon will result. Eventually, the other loop also probably will be abandoned. How wonderful it would seem to be present at the proper moment to witness such an event particularly if one had a time-lapse movie camera to record it for posterity! (**Figure 91**).

About a mile and a half below the south saddle of The Loop, the mouth of Salt Creek occurs which drains a large part of the Needles district. Salt Creek Canyon is about 2 airline miles above the mouth looking southeast toward Six-Shooter Peaks and Shay Mountain, northernmost of the Abajo Mountains on the horizon.

A mile and a half above the confluence is The Slide, a jumbled mass of angular blocks of rock that fell from the northwest canyon wall and originally probably extended all the way to the southeast bank of the river. It still extends nearly across the river leaving only a narrow deep chute along the southeast bank. Rough fast water occurs in the chute with waves about 2 feet high. At higher stages of the river, progressively more of The Slide is covered by water, and there is less of a tendency for waves to form. The date of this landslide is not known, but it may well have occurred during prehistoric times (**Figure 92**).

Figure 91. The Loop West (right) is still in the process of developing a rincon but not as fully developed as The Loop East which occurs at a lower elevation. The arrow points to a thrust fault which can be traced into the western Loop saddle.

Soon we reach the confluence of the Green and Colorado Rivers. This important junction of two mighty rivers was noted by all previous voyagers, but their impressions of it differed considerably.

Figure 92. The Slide is positioned about 1.5 miles north of the Green River-Colorado River confluence on the Colorado River.

Cataract Canyon heads at the confluence, but the rapids do not appear until we leave Spanish Bottom some 3.5 miles below. Between The Loop and Spanish Bottom, the Colorado River follows closely the axis of an anticline. Along this reach the rock strata dip downward away from the river. This structure was designated as the Meander anticline. The weight of the rocks on each side of the river had squeezed underlying beds of salt in the Paradox Member of the Hermosa Formation and caused them to move upward along the river where the confining strata had been removed by erosion. Thus we have what may be termed an erosional anticline whose axis, or crest, follows the river. Erosional anticlines also occur elsewhere as along the Eagle and Roaring Fork valleys of central Colorado. This zone of weakness in Canyonlands overlies and follows a break in the hard Precambrian rocks that underlie the area at great depth.

Smooth water continues from the confluence to Spanish Bottom where the Old Spanish Trail comes down to the river from the west and continues up Lower Red Lake Canyon to the east. This is about the south end of the Meander anticline, and an intruded chunk of the Paradox Member, mostly gypsum occupies part of the mouth of Lower Red Lake Canyon (**Figure 93**).

The remaining 10 miles or so of Cataract Canyon within Canyonlands National Park contains many rapids and should be traversed only under the leadership of experienced river guides. If and when Lake Powell reaches its maximum level, it will extend to within about a mile of the park, but at present it heads near the mouth of Gypsum Canyon, about 5 miles below the park.

Chapter 5. Geologic History Summary

Having finished our geologic ramble through Canyonlands National Park, this pile of eroded rocks fit into the bigger scheme of things, the geologic age and events of the earth as a whole. The rock strata still preserved in the park ranges in age from Pennsylvanian to Jurassic, or from about 300 to 175 million years ago, a span of about 125 million years. This seems an incredibly long time until it is noted that the earth is some 4.5 billion years old and that our rock pile is but one twenty-fifth, or 4 percent of the age of the earth as a whole. But this is not the whole story. As indicated earlier, about 10,000 feet of younger Mesozoic and Tertiary rocks that once covered the area have been carried away by erosion, and if these are included in the stratigraphic pile of rocks, the span is increased to about 250 million years.

Figure 93. The lower part of the Red Lake Canyon exposes gypsum in the lower part of the canyon (gray beds below the red bed cliffs).

Deep tests for oil and gas tell us that much older rocks underlie the area, and that some of these rocks played a part in shaping the park noted today. The breaks in the deep-seated Precambrian rocks and the salt in the Paradox Member represent periods of non-deposition and erosion of the upper surfaces. In addition to the Precambrian igneous and metamorphic rocks, there are about 2,000 feet of Paleozoic sedimentary rocks older than the Pennsylvanian Paradox Member. Most of these sedimentary rocks were laid down in ancient seas during Cambrian, Ordovician, Devonian, Mississippian, and Pennsylvanian times.

There are some gaps in the rock record caused by temporary emergence of the land above sea level and erosion of the land surface before the land again subsided below sea level so that deposition could resume. Silurian rocks are absent altogether, presumably because here the Silurian Period was dominated by erosion rather than deposition.

While Pennsylvanian and Permian sediments were being deposited in and southwest of the park, a large area to the northeast was called by geologists the Uncompahgre highland, because it occupied the same general area as the present Uncompahgre Plateau which rose slowly above sea level. Whatever Paleozoic rocks there were on this rising land, part of the underlying Precambrian rocks were eroded and carried by streams into deep basins to the northeast and southwest. Thus, while mostly marine or near shore deposits were being laid down in and near the park, thousands of feet of red beds were being laid down by streams in an area between the park and the Uncompahgre Plateau.

During part of Middle Pennsylvanian time, a large area including the park known as the Paradox Basin was alternately connected to or cut off from the sea so the water evaporated during cutoff periods and was replenished during periods when connection with the sea resumed. In this huge evaporation basin were deposited the layers of salt and gypsum plus some potash salts and shale that now make up the Paradox Member. The old Uncompahgre highland continued to shed debris into the bordering basins until Triassic time when it began to acquire a veneer of red sandstone and siltstone of the Chinle Formation. The area remained above sea level during the Triassic Period and most if not all the Jurassic Period apart from the Jurassic Carmel Formation was laid down in a sea that lay just to the west.

Late in the Cretaceous Period, a large part of central and southeastern United States including the eastern half of Utah sank beneath the sea, receiving thousands of feet of mud, silt, and some sand that later compacted into the Mancos Shale. This formation and all the younger and some older strata have long since been eroded from the park area but are present in adjacent areas such as the lower slopes of the Book Cliffs north of Green River, Crescent Junction, and Cisco. The land rose above the sea at about the close of the Cretaceous and has remained above ever since, although inland basins and lakes received sediment during parts of the Tertiary Period.

Compressive forces in the earth's crust produced some gentle folding of the strata at the close of the Cretaceous, but more pronounced folding and some faulting occurred during the Eocene Epoch when most of the Rocky Mountains took form. During the Miocene Epoch, molten igneous rock welled up into the strata to form the cores of the nearby La Sal, Abajo, and Henry Mountains. Additional uplift and some folding occurred in the Pliocene and Pleistocene Epochs.

Much of the course of the Colorado River was established in the Miocene Epoch with some additional adjustments in the late Pliocene and early Pleistocene Epochs. Erosion during much of the Tertiary Period and all of the Quaternary Period combined with some sagging and breaking of the crust brought on by solution and lateral squeezing of salt beds beneath The Needles, The Grabens, and the Meander anticline producing the landscape as we now see it. The Precambrian rocks beneath the area are about 1.5 billion years old, so an enormous span of time is represented by the rocks and events in and beneath Canyonlands National Park.

Considering the geologic formations that make up the Colorado Plateau including national parks (N.P.), national monuments (N.M.) (excluding small historical or archeological ones), Monument Valley, San Rafael Swell, and Glen Canyon National Recreation Area, certain formations or groups of formations play starring roles in some parks or monuments, some play supporting roles and in a few places the entire cast of rocks gets about equal billing. For example, Dinosaur N. M. with exposed rocks ranging in age from Precambrian to Cretaceous represents the greatest time span (nearly 2 billion years) but has one unit-the Jurassic Morrison Formation. This unit contains the many dinosaur fossils that give the monument its name and fame. Several older units have supporting roles. Grand Canyon N. P. and N. M. are next with rocks from Precambrian through Permian excluding the Quaternary lava flows in the N. M., but here there is truly a team effort for the entire cast gets about equal billing.

Canyonlands N. P. stands third in size of cast with rocks ranging from Pennsylvanian to Jurassic, but top billing would fall to the Permian Cedar Mesa Sandstone Member of the Cutler Formation from which The Needles, The Grabens, and most of the arches were sculptured. Triassic Wingate Sandstone and Kayenta Formation get second billing for their roles in forming and preserving Island in the Sky and other high mesas.

Considering those with only one or few players in the cast, Black Canyon of the Gunnison N. M. cut entirely in rocks of early Precambrian age (except for only a veneer of much younger rocks) obviously has but one star in its cast. Colorado N. M. contains rocks ranging from Precambrian to Cretaceous-equal to Dinosaur in this respect it is unique in that all the rocks of the long Paleozoic Era and some others are missing from the cast. Of those that remain, the Triassic Wingate and Kayenta are the stars with strong support from the Jurassic Entrada Sandstone.

All the bridges in Natural Bridges N. M. were carved from the Permian Cedar Mesa Sandstone, also the star in Canyonlands N.P. In Canyon de Chelly (pronounced "dee shay") N. M. and Monument Valley (neither N. P. nor N. M., as it is owned and administered by the Navajo Tribe), the de Chelly Sandstone Member of the Cutler Formation, a Permian member younger than the Cedar Mesa-plays the starring role.

Wupatki N. M., near Flagstaff, Ariz. stars the Triassic Moenkopi Formation. Petrified Forest N. P. (which now includes part of the Painted Desert) also has but one star: the Triassic Chinle Formation with its many petrified logs and stumps of ancient trees. The Triassic-Jurassic Glen Canyon Group which includes the Triassic Wingate Sandstone and Kayenta Formation and the Triassic-Jurassic Navajo Sandstone receives top billing in recently enlarged Capitol Reef N. P., but the Triassic Moenkopi and Chinle Formations enjoy supporting roles.

The Triassic-Jurassic Navajo Sandstone, erosional remnants of which are found on the high mesas of Canyonlands N. P. is the undisputed star of Zion N. P., Rainbow Bridge N. M., and Glen Canyon National Recreation Area despite the fact that the latter is the type locality of the entire Glen Canyon Group. The Navajo also forms the impressive reef at the eastern edge of the beautiful San Rafael Swell (a dome, or closed anticline) now crossed by Highway 1-70 between Green River and Fremont Junction, Utah.

Continuing upward in time, the Jurassic Entrada Sandstone which stars in Arches N. P. with help from the underlying Navajo Sandstone and a supporting cast of both older and younger rocks is encountered. The Entrada also forms the grotesque erosional forms called "hoodoos and goblins" in Goblin Valley State Park, north of Hanksville, Utah. Moving ever upward in time, the Cretaceous Mesaverde Group whose caves in Mesa Verde N. P. house beautifully preserved ruins once occupied by the Anasazi, the same ancient people who once dwelt in Canyonlands N. P.

This brings us up to the Tertiary Period, during the early part of which the pink limestones and shales of the Paleocene and Eocene Wasatch Formation were laid down in inland basins. Beautifully sculptured cliffs, pinnacles, and caves of the Wasatch star in Bryce Canyon N. P. and nearby Cedar Breaks N. M. This concludes the geologic timing except for Quaternary volcanoes and some older volcanic features at Sunset Crater N. M., near Flagstaff, Ariz. Thus, one way or another many geologic units that formed during the last couple of billion years have performed on the stage of the Colorado Plateau still lurk in the wings eagerly awaiting applause to recall them to the footlights. Don't let them down-visit and enjoy the national parks and monuments of the Plateau, for they probably are the greatest collection of scenic wonderlands in the world.

References

Lohamn, S.W. 1974. The geologic story of Canyonlands National Park. Geological Survey Bulletin 1327, United States Geological Survey including references contained within.

Wikipedia, 2021. Canyonlands National Park. Posted on the internet.

www.ingramcontent.com/pod-product-compliance
Lightning Source LLC
Chambersburg PA
CBHW040221220526
45473CB00001B/68